Handbook
of
Carpentry
and
Joinery

CARPENTER'S WORK IN THE EARLY STAGE OF HOUSE BUILDING

Carpentry begins soon after the foundations have been laid. Once the damp-proof course is in position the door frames can be erected.

Handbook
of
Carpentry
and
Joinery

A.B. Emary

 Sterling Publishing Co., Inc. New York

Library of Congress Cataloging in Publication Data

Emary, A. B. (Alfred Bethuen)
  Handbook of carpentry and joinery.

  (Home craftsman series)
  Combined ed. of two books, Practical carpentry and
joinery, c1963 and Advanced carpentry and joinery,
c1967.
  Includes index.
  1. Carpentry—Amateurs' manuals. 2. Joinery—
Amateurs' manuals. I. Emary, A. B. (Alfred Bethuen)
Advanced carpentry and joinery. 1981. II. Title.
III. Series.
TH5606.E53 1981   694   81-50984
ISBN 0-8069-7536-9 (pbk.)   AACR2

Published 1981 by Sterling Publishing Company, Inc.
Two Park Avenue, New York, N.Y. 10016

Available in Canada from Oak Tree Press, Ltd.
℅ Canadian Manda Group, 215 Lakeshore Boulevard   East,
Toronto, Ontario M5A 3W9
Manufactured in the United States of America
Library of Congress Catalog Card No.: 81-50984
Sterling ISBN 0-8069-7536-9 Paper

Contents

Preface 6

Practical Carpentry and Joinery

1. Timber—The Ideal Material 7
2. Tools 23
3. Common Joints 39
4. Fixing Devices 49
5. Craft Geometry 59
6. Setting-out Rods 74
7. Marking Out the Material 78
8. Formwork for Concrete 85
9. Centres for Arch Work 94
10. Ground Floors 101
11. Upper Floors 105
12. Roof Construction 114
13. Stud and Glazed Partitions 132
14. Windows 136
15. Doors and Door Frames 149
16. Splayed Work 162
17. Staircase Work 168
18. Shoring 180

Advanced Carpentry and Joinery

19. Timbering to Excavations 196
20. Centres for Arches 199
21. Gantries 204
22. Formwork for Concrete 207
23. Roofs 214
24. Geometry and the Steel Square in Roofing 225
25. Doors 240
26. Sliding Door Gear 249
27. Windows with Curved Heads 254
28. Roof Lights and Ventilators 257
29. Panelling to Walls 269
30. Counter Construction 274
31. The Construction of Stairs and Handrails 276
32. Glulam Work 300
33. Applied Geometry 308

INTRODUCTION

THE ANCIENT civilizations had many skilled woodworkers and many examples of their skill still survive today. The carpenter was the designer and builder of the homes of the people and many found employment on those marvels of craftsmanship, the Gothic cathedrals. The first clocks, the first telescope, the first printing press were made of wood by the carpenter. In those times the only way of becoming a carpenter was to serve an apprenticeship with a master carpenter. The secrets of the trade were not written down but imparted to the apprentice during the time he was bound to serve his master.

In the 16th century, the craft of joiner emerged, primarily due to the invention of the panelled frame. Quarrels between the Guild of Carpenters and the newly formed Guild of Joiners resulted in joiners' work being defined as "all formes of fframes made with mortisses and tennants glewed, pynned or duftalled . . . all sorts of work made for ornament or beautie which cannott bee made without the use of Glew".

Within recent times, the tempo of life demands more concentrated training during a shorter period of apprenticeship. This book has been written to help the novice acquire that basic knowledge so essential to the carpenter and joiner of today. This handbook has been compiled from the author's book on intermediate carpentry and its companion volume on advanced carpentry and joinery.

Although many of the finer parts of the crafts of carpentry and joinery seem to be disappearing today for economic reasons and also what appears to be a lack of interest, the author feels that there will always be a large percentage of craftsmen who will still keep that pride of craftsmanship within them and who will always be able to 'turn out' a good job of work when called upon to do so.

CHAPTER I TIMBER—THE IDEAL MATERIAL

A GOOD KNOWLEDGE of the materials one uses to obtain a livelihood is necessary to any craftsman. With this knowledge he is able to select the right material for any specific purpose. There are many hundreds of different timbers which are used for commercial purposes and, provided each of these is used for work to which it is best suited, timber can be regarded as the ideal material.

Varied uses of timbers. It can be used for almost any purpose, a few of which are the building of houses, bridges, and aeroplanes; it can be used for fuel to obtain warmth and power; the manufacture of furniture for our comfort; extracts from woods assist in the making of medicines; and in certain circumstances trees also provide food.

How does a tree grow? A seed from a tree drops to the ground and, if conditions are favourable, germination takes place. The seed sends a root downwards to obtain the moisture from the ground, and a small shoot grows upwards to produce leaves. The moisture travels upwards to the leaves, and these change the water, or sap, into food. Each growing season the young tree will put on a layer of new cells which completely cover the cells produced the previous season. In addition to the new cells being produced, the buds at the ends of the stem and branches throw out additional shoots, and so the sapling develops into a mature tree (Fig. 1).

Components of a tree. There are three main parts of a tree; the roots, the stem or trunk, and the leaves (Fig. 2). The roots absorb moisture from the soil, and this moisture travels upwards through the sapwood to the leaves, where, through the action of the sun, the moisture or sap, as it is commonly called, is changed into sugars and starches which are returned to the stem to act as food for the living tree.

Softwoods and hardwoods. Commercial timbers are divided into two groups, namely, softwoods and hardwoods. Generally speaking the softwoods are the cone-bearing trees which have needle-like leaves and are usually evergreen (Fig. 3). Hardwoods are the broad-leafed trees, and are deciduous, which means that the leaves fall to the ground at the end of each growing season (Fig. 4). There are exceptions, of course, the holly tree being one of these. Although a hardwood tree, it is evergreen and has leaves throughout the twelve months of the year.

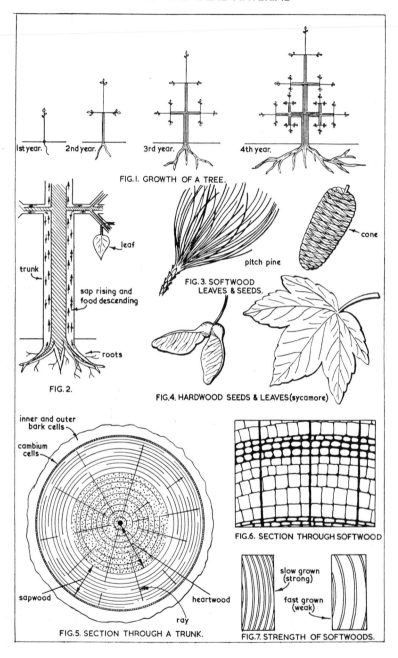

1st year.   2nd year.   3rd year.   4th year.

FIG.I. GROWTH OF A TREE.

leaf

trunk

sap rising and food descending

roots

FIG. 2.

pitch pine

cone

FIG.3. SOFTWOOD LEAVES & SEEDS.

FIG.4. HARDWOOD SEEDS & LEAVES (sycamore)

inner and outer bark cells

cambium cells

sapwood

heartwood

ray

FIG.5. SECTION THROUGH A TRUNK.

FIG.6. SECTION THROUGH SOFTWOOD

slow grown (strong)

fast grown (weak)

FIG.7. STRENGTH OF SOFTWOODS.

The seeds of hardwood trees are to be found in the fruit which the trees produce. For instance the seeds of the holly tree are in the red berries which are in such great demand at Christmas time for decorative purposes. The seeds of the horse-chestnut tree are to be found in the fruit which are commonly called "conkers". The seeds, it will be realised, are always enclosed in a case or covering. The seeds of softwood trees are in the cones which the trees produce.

When the seeds are ripe the cone opens and leaves the seeds exposed, and they are scattered by the winds as well as by just falling to the ground.

Parts of a tree. If the end of a tree-trunk which has been recently felled is examined, the appearance would probably be that shown in Fig. 5. This is a horizontal section through a tree. The outer surface of the trunk is protected by the bark or bast. Immediately below the bark is a layer of cells which produce the new wood and bark cells each season. These are called cambium cells. They produce new wood cells inwardly, and new bark cells outwardly. Below the cambium cells is the sapwood and often the sapwood area is of a lighter colour than that of the middle portion of the trunk. Most of the water the roots pick up from the ground travels up through the sapwood on its way to the leaves.

The darker or central portion of the trunk is the more mature wood and is called heartwood. There is often a marked difference in colour between sapwood and heartwood, and many architects when specifying for a high-class joinery contract will mention that heartwood only is to be used. It is the different colours of the two woods which bar the use of sapwood in many instances. The difference in strength is small, but sapwood is more prone to attack by fungi or beetles because it contains more foodstuff on which these attackers live. This alone should not prevent the use of sapwood, because, properly treated, it can be made immune from such attacks. In the centre of the trunk is the pith or medulla, and is the first portion of the growth at that particular part of the trunk.

Growth rings. As already mentioned a tree puts on a layer of new cells each growing season, and if the end of the trunk is looked at, these cells, or growth rings as they are called, can be clearly seen in many timbers. The first part of each growth ring consists of springwood cells, and, as the name implies, were grown in the early part of the season. These cells are thin walled and usually light in colour. The second half of the growth ring consists of cells with rather thicker walls called summerwood cells. They are much darker than the spring growth.

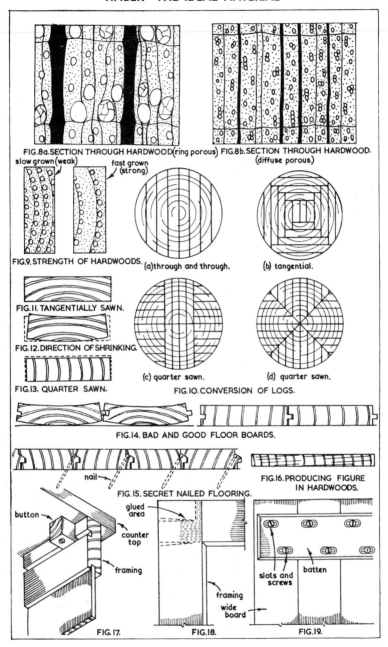

FIG.8a. SECTION THROUGH HARDWOOD(ring porous) FIG.8b. SECTION THROUGH HARDWOOD.
(diffuse porous.)

slow grown (weak)   fast grown (strong)

FIG.9. STRENGTH OF HARDWOODS. (a) through and through.   (b) tangential.

FIG.11. TANGENTIALLY SAWN.

FIG.12. DIRECTION OF SHRINKING.

FIG.13. QUARTER SAWN.   (c) quarter sawn.   (d) quarter sawn.

FIG.10. CONVERSION OF LOGS.

FIG.14. BAD AND GOOD FLOOR BOARDS.

nail

FIG.15. SECRET NAILED FLOORING.

FIG.16. PRODUCING FIGURE IN HARDWOODS.

button

glued area

counter top

framing

framing
wide board

slots and screws   batten

FIG.17.   FIG.18.   FIG.19.

The foregoing description of growth rings is in very general terms because there are many timbers, especially in the hardwoods, where the difference between spring and summerwood growths is small indeed, and no difference can be seen with the naked eye.

Horizontal rays. In all woods, too, there exist rays which are in the form of radiating lines of cells as seen in Fig. 5. Although these exist in all timbers, in many instances they cannot be seen without the help of a lens or a microscope because they are so narrow. In other timbers, such as oak or beech, these rays can be seen quite easily with the naked eye and, in fact, it is these rays which give such timbers as oak and London plane their very rich figure if used to advantage. This will be discussed later.

Growth cells. If a closer look is made with a microscope at the end-grain surfaces of a piece of timber much more detail of the parts which make a piece of wood will be seen. Take a piece of softwood for the first example (Fig. 6). This shows that the piece of wood consists mainly of hollow tubes called tracheids. The thin-walled cells were produced in the spring and the thicker-walled cells in the summer and autumn. It is the summerwood or thick-walled tracheids which give softwoods their strength, and so it follows that any timber with thick-walled tracheids and which has been grown slowly will have more growth rings per inch and so will be a much stronger timber than one which has been growing quickly (Fig. 7).

The lines of cells which run from top to bottom of the drawing are the rays or horizontal parenchyma cells and are where food is stored for the living tree. In softwoods these are usually too small to be seen with the naked eye, but they are most important for identification purposes. Many softwoods can be identified by a close study of the horizontal parenchyma cells.

There are two types of hardwoods namely ring-porous and diffuse-porous (Fig. 8). A ring-porous timber is illustrated in *a* (Fig. 8), which is a view of the end grain of a piece of English oak. Hardwoods consist mainly of vessels, fibres, and rays of horizontal parenchyma cells. In ring-porous timbers rather large vessels are produced during the early part of the growth ring and as the season progresses the size of the vessels get smaller.

In diffuse-porous timbers, however, the size of the vessels are by and large the same size throughout the width of each growth ring and are usually scattered fairly evenly over the area. Fig. 8*b* is a view of the end grain of a piece of sycamore, which is diffuse porous.

The space between the vessels is filled mainly with fibres and it is these which give strength to a hardwood tree (Fig. 9).

If a ring-porous hardwood is slow-grown there are many more

large vessels per inch in width and less fibres, so a much weaker timber is produced when compared with a fast-grown hardwood tree. The thick, black lines running from top to bottom of Fig. 8 are the rays or horizontal parenchyma cells, and in hardwoods these can often be seen with the naked eye. The thin lines are also rays. It is the horizontal rays which give figure to many hardwood timbers.

The best time for felling trees is in the autumn or winter. This is because the atmosphere is usually damp, and so shrinkage splits and shakes are less likely to arise. In addition to this the attacks of a fungus or beetles are much less likely to occur during cold weather.

Conversion. When a tree has been felled the sooner it is transported to a timber merchant for converting into smaller pieces the better. A small piece of timber is much easier to dry than a large piece. There are several ways of cutting up a log, and some of these are illustrated in Fig. 10. The first is through-and-through conversion, and consists solely of passing the log backwards and forwards over a saw-bench to produce a series of boards of any required thickness (Fig. 10a). If boards are required for specific purposes then the logs should be converted in a particular way.

It has already been mentioned that softwoods obtain their strength from the growth rings. Consequently for floor-joists, which must have maximum strength to span between two walls, the growth rings should travel through the wide dimension of the joists as in Fig. 11. It follows then that if a softwood log is to be converted to produce strong lengths of timber the boards must be tangentially sawn (Fig. 10b). This consists of passing the log over the saw-bench and turning the log over on to an adjacent side after each cut. All the boards produced in this way will be tangentially sawn. Boards tangentially sawn will shrink a great deal when they are dried. If not dried sufficiently they will shrink again when they are fixed in a warm building. The direction of most of the shrinkage is in the direction of the growth rings (see Fig. 12).

Boards for joinery purposes, and where strength is of secondary importance, should be produced by cutting up the log in one of the ways illustrated in Figs. 10c and d. This is quarter-sawing, and all the boards will have the growth rings running through their thickness or shortest dimension. The shrinkage of these boards will have been reduced to a minimum because most of the shrinkage takes place in the direction of the growth rings, and when boards are quarter-sawn the direction of the growth ring is through the smallest dimension of the board, Fig. 13.

Floorboard shrinkage. Fig. 14 illustrates two types of floor-boards and what happens when they shrink. The first shows tangentially sawn boards which have dried considerably since they were fixed. As shrinkage takes place in the direction of the growth rings, not only do wide gaps show between the boards, but most of them have also cupped. The corners which point upwards will show through any lino which is laid down. The second drawing shows flooring which has been quarter-sawn and little shrinkage has taken place. The ideal flooring, especially for good-class work, is narrow, quarter-sawn, secret-nailed boards as seen in the drawing (Fig. 15).

When timbers, such as English oak, beech, Australian silky oak, etc., have wide horizontal rays, the value of the timber, from a decorative point of view, can be greatly increased by quarter-sawing the boards. This ensures that the rays will strike through the wide surfaces of the boards as shown in Fig. 16. Not only are boards obtained which will not shrink to any great extent but the value of the timber is increased because of its fine figure.

SEASONING

When a tree is felled it contains a vast amount of water within its cells. The moisture not only saturates the cell walls but also fills the cell cavities. To make the timber more suitable for use most of the water has to be removed, and the process of this is called seasoning. When timber begins to dry the moisture begins to move towards the surfaces where it is evaporated into the air. If drying is continued after the water in the cell cavities has been removed, the timber will begin to shrink. This shrinkage continues until the timber ceases to give off any more water. If it starts to absorb more water then the timber starts to swell again.

When timber is allowed to dry naturally it will give off moisture until it is in equilibrium with the moisture in the surrounding atmosphere. On the other hand, dry, unprotected timber will absorb moisture if the conditions around which it is fixed are damp. That is why unprotected timber in the shape of doors or windows, or what have you, will automatically dry out during a dry hot summer, and then as the wet days of autumn and winter arrive, will absorb moisture and swell to such an extent as to possibly need easing with a plane to make them work properly. Even when protected, small amounts of moisture will enter timber through the protection and it is necessary, especially where wide boards are used, to use certain joints and methods which will allow the timber to shrink and swell without restriction (see Figs. 17, 18, 19, 20).

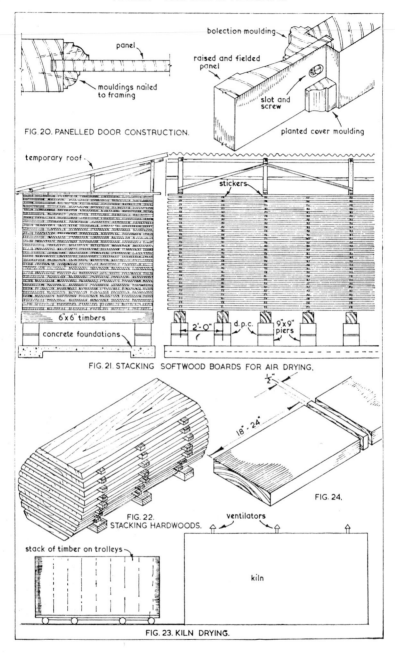

FIG. 20. PANELLED DOOR CONSTRUCTION.

panel

mouldings nailed to framing

bolection moulding

raised and fielded panel

slot and screw

planted cover moulding

temporary roof

stickers

6"x6" timbers

concrete foundations

2'-0"   d.p.c.   9"x9" piers

FIG. 21. STACKING SOFTWOOD BOARDS FOR AIR DRYING.

½"

18" - 24"

FIG. 24.

FIG. 22.
STACKING HARDWOODS.

ventilators

stack of timber on trolleys

kiln

FIG. 23. KILN DRYING.

It is obvious, then, that once a piece of timber has been dried to suit the particular surrounding in which it is to be fixed, it must be protected to avoid as far as possible the inevitable swelling and bad fitting which will be the result. There are two methods of drying timber; air drying and kiln drying.

Air seasoning. This consists of stacking the planks in a correct fashion in a position where there is adequate ventilation but where no direct moisture or rain can settle on the planks (see Fig. 21). This drawing shows a stack of softwood boards stacked in the open, well clear of the ground, and with a temporary roof over them to stop the rain from keeping the boards wet.

The idea is to get a flow of fresh air to pass over the four surfaces of all the boards so that the moisture contained in the boards is evaporated into the air. To do this, stickers or piling sticks must be placed between each layer of boards in vertical alignment and about 2 ft. apart. The stickers should be of the same type of timber as the boards and should be about $\frac{3}{4}$ in. thick and 1 in. wide.

The stack should not be more than 6 ft. wide, to ensure the centre boards getting an adequate amount of air passing over their surfaces, and not more than 10 ft. high. The ground around the stack should be well clear of rubbish, weeds, and other pieces of wood. If possible the site for the stack should be concreted over. Short 9 in. by 9 in. brick piers—about 12–18 in. high should be constructed about 2 ft. apart as seen in the drawing, and bitumen damp-proof courses incorporated in them at least 6 in. from ground level to prevent rising dampness from reaching the timber.

Heavy timbers, as seen in the drawing, form a foundation for the boards. A space of about 1 in. is left between the boards in each layer to allow for an upwards movement of the air currents. A pitched roof of corrugated iron or asbestos which overhangs adequately on each side is secured on the top of the stack with ropes or some other means. Small amounts of hardwood can be stacked in a similar manner to that shown in Fig. 22 as the boards come off the saw.

Kiln seasoning. The drying of timber by the air-seasoning method is not sufficient to bring the moisture content down to much below 20%, even when the weather is dry and warm. To make timber more suitable for joinery purposes a second method must be used. This is called kiln seasoning. Large joinery firms have their own kilns, and when operated by an experienced man the moisture content in the timber can be brought down to any required level without any serious degrades resulting, such as splits, warping, etc.

A kiln consists of a brick building (Fig. 23) into which can be transported on trolleys, a carefully stacked load of timber. As in air drying, piling sticks are placed, about 2 ft. apart, between each layer of boards.

The air in the kiln can be heated by steam pipes and can be kept circulating through the stack by fans. These fans can be reversed

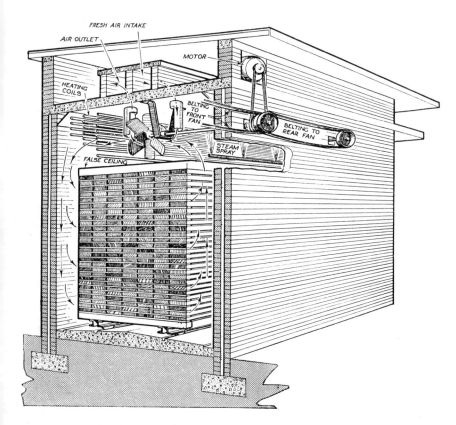

CROSS-SHAFT OVERHEAD FAN KILN
The arrows show the direction of the air circulation.
This drawing was prepared from plans kindly lent by the Forest Products Research Laboratory.

at any time to get the air circulating in the opposite direction. This action will enable the operator to keep the whole stack at a fairly even moisture content. So that the outsides of the boards will not dry out too quickly and so cause the boards to split on the surfaces, steam is injected into the kiln when the drying begins to keep the

MOISTURE CONTENTS OF TIMBER FOR VARIOUS PURPOSES

The Figures for Different Species vary, and the Chart shows only Average Values.

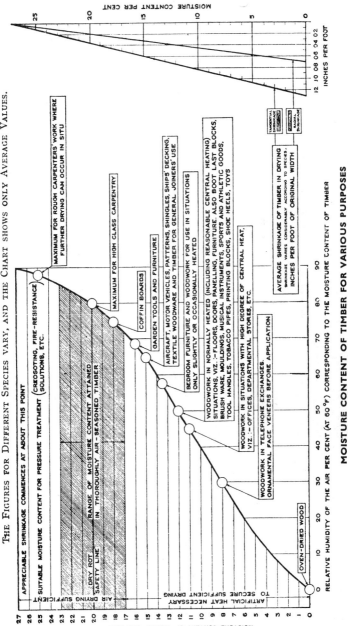

MOISTURE CONTENT OF TIMBER FOR VARIOUS PURPOSES

The Chart shows average values. The figures may vary for different species.

From Forest Products Research Laboratory, Leaflet No. 9, Crown copyright reserved. Reproduced by permission of the Controller, Her Majesty's Stationery Office.

17

humidity fairly high. This enables the moisture in the centres of the boards to move towards the surfaces while the surfaces themselves are kept fairly moist.

Humidity and heat ratio. Unless the humidity of the kiln is kept high at the beginning of the drying operation, the outside surfaces

BRICKWORK CARRIED UP TO DAMP-PROOF COURSE
The latter is essential to avoid saturation of timber.

will quickly give off the moisture they contain and then shrink, resulting in surface checks. In addition to these a condition known as case hardening would occur, and stresses would be set up in the boards. As the drying progresses the humidity in the kiln is lowered and the heat raised, in accordance with carefully worked out schedules, until the required moisture content is reached. The humidity and heat are checked with wet- and dry-bulb thermometers which are placed in various positions round the kiln.

The advantage of kiln seasoning over air seasoning is, of course, the fact that the moisture content of timber can be brought down to any level in a kiln, whereas in air seasoning it can rarely be brought down to a level below 18%. Time, too, is an important factor, kiln seasoning being much quicker of the two. As a precaution, in either method, the ends of the boards should be sealed with some solution to prevent the ends from drying too quickly. The end grain of a piece of timber will give up its moisture much more readily than any other portion of the board.

Ascertaining moisture content. The moisture content of a piece or a parcel of timber can be determined at any time by following this procedure. Select a board from the stack and cut a $\frac{1}{2}$ in. length from it about 2 ft. from one end (Fig. 24). It should be weighed on some accurate scales or balances and the weight noted. The sample should be placed in an oven with a temperature of a little more than 100° centigrade and reweighed at intervals until it is clearly losing no more weight. This last weight is the dry weight of the sample. To obtain the moisture content of the board, or the stack of timber at the time when the sample was taken from it, the following calculations should be made:

$$\frac{\text{Initial weight} - \text{dry weight}}{\text{dry weight}} \times \frac{100}{1} = \text{M.C. }\%$$

Supposing the initial weight of the sample is 48 gm. and the dry weight is 36 gm. the moisture content will be found thus:

$$\frac{48-36}{36} \times \frac{100}{1} = \frac{12}{36} \times \frac{100}{1} = 33\tfrac{1}{3}\% \text{ (of its dry weight)}$$

TIMBER DEFECTS

There are many kinds of defects which can affect the value and the uses of timber. Some of these are natural defects and have to be accepted, but some are brought about after felling has occurred and can and should be avoided.

Knots. The commonest of the natural defects are knots. They are caused by branches growing out from the tree and subsequently being enclosed by the layers which are grown each season. Small, live knots do not affect the strength of a piece of timber to a great extent, but the appearance is greatly affected. Large knots do affect the strength, and these must be taken into consideration when they are used for constructional purposes. If floor-joists which are supported at each end have large knots they should be placed in position with the knots in the upper half of the joist. This portion is in compression whereas the lower half is in tension. Cantilevered joists should, if possible, have the large knots in their lower halves, for the same reason, Fig. 27.

Shakes. Shakes or splits can greatly affect the economical use of a piece or timber or a log. Heart shakes and star shakes (Fig. 28) usually occur in the living tree and cannot be helped, but end shakes and surface checks (Fig. 29) which occur after felling can be avoided if the timber is quickly converted after felling and is carefully seasoned. Cup shakes are a natural defect and are due to the weakness of adhesion between one growth ring and another.

Bowing. Bowing (Fig. 30) and twisting (Fig. 31) are two more defects which can render a board useless in its original length, and can be blamed on bad drying.

Case hardening, which has already been mentioned, is another seasoning defect and which can and should be avoided.

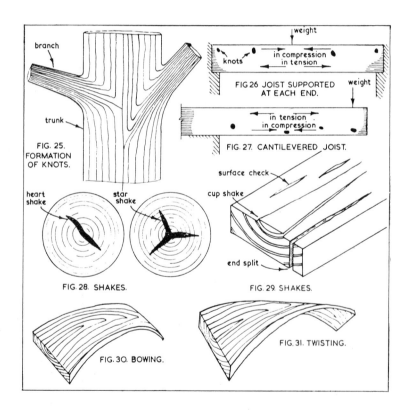

FIG. 25. FORMATION OF KNOTS.

FIG 26 JOIST SUPPORTED AT EACH END.

FIG. 27. CANTILEVERED JOIST.

FIG. 28. SHAKES.

FIG. 29. SHAKES.

FIG. 30. BOWING.

FIG. 31. TWISTING.

SOFTWOODS

Name	Botanical Name	Obtained From	Weight Seasoned lbs./cu. ft.	Durability	Resistance to Preservation	Working Characteristics
Douglas Fir Oregon Pine British Columbian Pine	*Pseudotsuga taxifolia*	U.S.A. Canada Britain	32–35 lb.	Moderate	Fairly resistant	Tends to split when nailed Glues well Planes well
European Redwood Red or Yellow Deal Scots Pine Redwood	*Pinus sylvestris*	Northern Europe Russia Britain	30–33 lb.	Not durable	Only slightly resistant	Works well Nails well Glues well
Western Hemlock Pacific Hemlock British Columbian Hemlock	*Tsuga heterophylla*	Canada U.S.A. Alaska	30–32 lb.	Not durable	Fairly resistant	More difficult to work than European redwood Tends to split when nailed Glues well
Whitewood White Deal Baltic Whitewood European Spruce White Fir	*Picea abies*	Northern and Central Europe Britain	28–30 lb.	Not durable	Fairly resistant	Works well Nails well Glues well

21

TIMBER—THE IDEAL MATERIAL

HARDWOODS

Name	Obtained From	Weight lbs./cu. ft.	Durability	Working Characteristics
English Oak (*Quercus robor*)	Gt. Britain Europe	46–48 lbs.	Very durable	Difficult to work, glues well. Nails and screws well
Honduras Mahogany (*Swietenia macrophylla*)	British Honduras Central America	33–35 lbs.	Fairly good	Very good. Nails and screws well. Glues well
African Mahogany (*Ivorensis*)	West Africa	35 lbs.	Moderately resistant to decay	Difficult owing to interlocked grain
Elm (*Ulmus procera*)	Gt. Britain Parts Europe	33–35 lbs.	Fairly resistant to decay	Difficult owing to twisted grain
Beech (*Fagus sylvatica*)	Europe Asia	45 lbs.	Not resistant to decay	Good. Bends well

CHAPTER 2 TOOLS

THE PURPOSE OF this chapter is not to teach the newcomer to carpentry and joinery how to become proficient in the use of tools, as the author is of the opinion that it is impossible for a book to do just that.   Rather it is to help the young craftsman in collecting a kit of tools and to assist him in choosing the right tool for each operation.

Cheap tools should never be bought, because they can never be relied upon to do the job properly.   Chisels with cheap handles which split easily; screwdrivers which become loose in the handles; plane irons which can never be sharpened to a razor edge; these are all signs of poor-quality tools.

Tools which have names stamped on them such as STANLEY, RECORD, SPEAR & JACKSON, MARPLES, etc. are ones which can be relied on to give long and faithful service so long as they are looked after, and used for the work they are intended for.

Saws.   The first saw most young carpenters and joiners obtain when entering the crafts is the hand saw.   Its correct name is the cross-cut saw, Fig. 1, and is similar in shape to the rip saw and panel saw.   It can be used for many purposes such as cutting timber to length, and is often used for ripping timber, that is, cutting down the lengths of boards in the direction of the grain.

This operation is really the job for the rip saw, though many craftsmen these days do not purchase this tool because most of the ripping is done by machine, and any ripping which has to be done by hand can be done with the cross-cut.   The main difference between these tools is in the shapes of the teeth.   The cross-cut has up to eight points to the inch and the teeth are shaped as in Fig. 2, whereas the rip saw has up to five points per inch with teeth shaped as in Fig. 3.

The panel saw is of similar shape to that in Fig. 1 but is a much smaller and lighter saw.   Its teeth are shaped in the same way as the cross-cut, and total about ten points to an inch.   This is often termed a bench tool and is used for small ripping jobs, cutting wedges and shoulders to wide tenons, etc.

The tenon saw, Fig. 4, is another cross-cut saw, but is used for much finer work such as the cutting of shoulders and the cheeks to tenons, recessing across the direction of the grain, mitreing, etc.

23

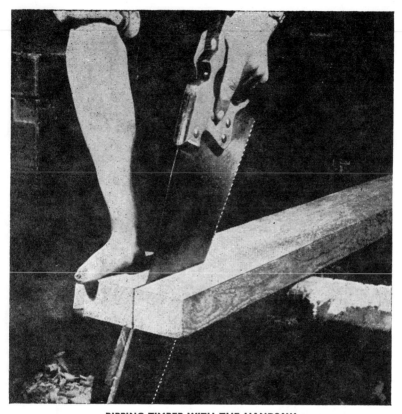

RIPPING TIMBER WITH THE HANDSAW
Carpenters mostly use the cross-cut saw only as this can be used for both ripping and cross-cutting.   Note index finger pointing along handle.

Its teeth are also similar to those in Fig. 2, but are up to twelve points to 1 in.   The tenon saw is fitted with a brass or steel back to give strength and support for the thin blade.   Common lengths for the foregoing saws are: rip saw 26–28 in., cross-cut 24–26 in., panel 20–22 in., tenon 10–14 in.   Many joiners prefer to have two tenon saws to hand on a bench, one about 14 in. in length—the other no more than 10 in. and with a fine blade for dovetails etc.

Saw Sharpening.   The sharpening of saws is a skilled job and only after a great deal of practice can one sharpen saws correctly. The teeth of saws after many sharpenings can become badly irregular in shape, and the points become out of line.   Before sharpening a

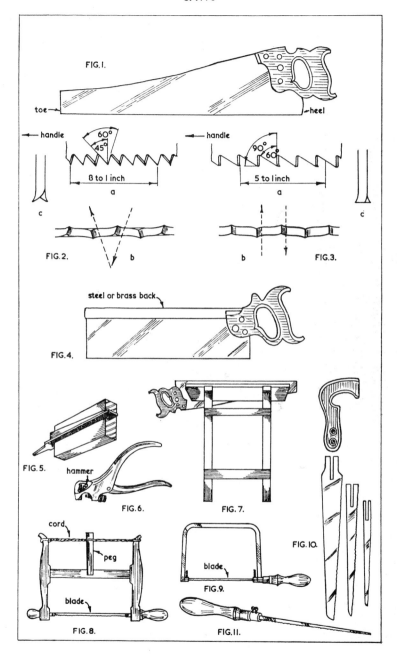

FIG. I.

toe ← ―heel

← handle   60° 45°   8 to 1 inch   a

c

FIG. 2.   b

b   FIG. 3.

← handle   90° 60°   5 to 1 inch   a

c

steel or brass back

FIG. 4.

FIG. 5.

hammer

FIG. 6.

FIG. 7.

cord   peg   blade   FIG. 8.

blade   FIG. 9.

FIG. 11.

FIG. 10.

25

saw in this state it is first necessary to run a flat file along the tops of all the teeth until the points are all in line with each other. The tool shown in Fig. 5 is used for this operation. It is merely a flat file wedged in a piece of wood which allows the operator to run the file along the tops of the teeth without damage to his hands.

The second operation in sharpening is to set the teeth. This means that every other tooth is bent slightly to one side of the blade and the others bent to the opposite side. To do this efficiently a saw setting tool is required, Fig. 6. This is adjustable according to the amount of set required. The saw set is placed on each alternate tooth and the handles squeezed together. The small hammer travels forward and pushes the tooth over to the correct amount of set. When each alternate tooth has been set the saw should be turned round and all the remaining teeth set in the opposite direction.

A saw vice, Fig. 7, is the best way of securing the saw during its sharpening. The jaws of the vice consist of two pieces of wood between which the saw is placed, and these fit into tapered recesses in the tops of the legs. When pressed downwards the recesses grip the jaws tightly and so prevent the saw from moving. A tap upwards at the ends of the jaws quickly releases the saw.

A three-cornered file, made specially for this work is held at each end by the hands, and with slight pressure downwards is allowed to travel throughout its length through the gullet of every alternate tooth. Two or three strokes per gullet is usually sufficient. The saw is then turned round again, and the remaining gullets filed. It must be remembered that the same pressure and the same number of strokes should be maintained throughout the sharpening.

For cross-cut saws the file should travel in the directions of the arrow shown in Fig. 2, and for rip saws in the direction indicated in Fig. 3. When sharpening cross-cut saws the file should be tilted upwards away from the operator, and the end of the file farthest away from the operator should always point towards the handle of the saw. With rip saws, the file is held almost level, and travels across the saw at right angles to the blade.

Saws for curved cuts. When sawing round curved lines a different saw is required. Figs. 8, 9, 10, and 11 illustrate four different saws in common use for this kind of work. All have very narrow or comparatively narrow blades. The bow saw, Fig. 8, is required for heavy work or where the timber to be cut is fairly thick. It has a wooden frame with a thin blade having a small hole at each end. The ends of the saw fit into slots attached to the handle spindles, and are held with rivets which pass through the holes. By turning the handles the cutting direction of the blade can be varied. Tension is placed on the blade by turning the peg which tightens up the cord fixed to the top ends of the frame.

The coping saw, Fig. 9, is similar to the bow saw, but the tension is obtained by the spring in the steel frame. The blade can be removed by turning the handle anti-clockwise which releases or produces the tension in the frame.

Fig. 10 shows a compass saw with three sizes of blades, and is used for fairly heavy work for which a bow saw is not practicable, such as cutting holes in flooring, etc. The blades are held in position by tightening the two screws seen in the handle. The keyhole saw, Fig. 11, is used for small jobs where a frame saw or compass saw could not operate. It obtains its name from the fact that it is useful for cutting the keyholes for locks.

Planes. There are many planes available to the craftsman these days, and to include them all in a chapter such as this would not be possible. Metal planes have largely taken the place of the wooden ones, but the old wooden jack plane, Fig. 12, is still as popular as ever, so it is felt that this one, at least, should be included. The jack plane is used for preparing the timber from the saw, in other words, its main function is to get the timber flat and out of winding, and to produce fairly reasonable surfaces prior to finishing with a smoothing plane.

The wooden jack plane is usually made of selected beech, and is about 17 in. long and 3 in. by 3 in. in cross section. An opening is made along the body of the tool to allow a cutting iron to be inserted and held in position by a wedge, see Figs. 12 and 13. A handle is fitted into the top surface near the back end of the plane. The iron passes through the depth of the plane, the opening being reduced at the lower surface sufficiently to allow only a small space (mouth) to remain between the cutting edge of the iron and the front edge of the mouth, Fig. 13.

Back iron. The cutting iron (Fig. 14a) has a hardened steel section at the cutting end which allows a sharp edge to be produced for preparing the timber. The back iron, Fig. 14a, is secured by means of a screw through the slot in the cutting iron. Its edge is close to the edge of the cutter, and its action is to break the shavings as they are raised from the timber, and so prevent tearing along the direction of the grain. The edge of the back iron should be a little less than $\frac{1}{8}$ in. from the cutting edge of the jack plane when preparing timber from the saw, and about $\frac{1}{16}$ in. or less when cleaning up timber with a smoothing plane.

A finished surface is not required from a jack plane, and so the cutting edge of its iron should be sharpened as in Fig. 15a. This will allow a fairly thick shaving to be removed easily whereas a smoothing plane iron should be shaped as in Fig. 15b. Its cutting edge should be straight with the corners removed. This will allow a good surface to be produced without the corners of the iron marking it.

Sharpening a plane. When sharpening a plane iron, two processes have to be carefully followed. First the iron has to be ground on a wet grindstone, the grinding angle about 25 degrees. The iron is then sharpened on an oilstone of medium or fine grade. The iron is tilted upwards slightly from the grinding angle. In theory the sharpening angle should be about 30–35 degrees, see Fig. 14b.

Wooden planes have largely been replaced by metal planes in recent years, and few craftsmen regard wooden planes as a good proposition. The metal ones are heavier than the wooden ones, which is a big point against them when they have to be carried regularly from job to job.

Metal planes. Figs. 16 and 18 illustrate a metal jack plane and a metal smoothing plane. The iron is held in position with a metal wedge seen in Fig. 16. Both are easily adjusted for depth of cut with the circular nut seen just in front of the handle, and laterally by the lever seen just behind the top of the cutting iron. The cutting irons and back irons for metal planes of the type illustrated in Figs. 16 and 18 are much thinner than those for wooden planes. Fig. 17 shows the irons for a metal plane.

Fig. 19 is a metal block plane which can be used for many bench jobs such as trimming mitres, removing arrises, cleaning up end grain of hardwoods, etc. The mouth on some block planes can be easily adjusted, and the iron can be adjusted in a similar manner as for the jack and smoothing planes.

The router, Fig. 20, is used for cleaning out the bottoms of grooves which run at right angles to the direction of the grain, such as housings for shelves, etc. The iron can be adjusted for depth by the circular nut seen behind the top of the cutting iron.

Rebate and grooving planes. The metal rebate plane seen in Fig. 21 is a great improvement on the old wooden type of rebate plane. It has two positions for the cutting iron, the front one making it possible to reach near to the end of a stopped rebate or similar sinking. An adjustable fence makes the cutting of a rebate a simple task, and a spur and a depth gauge on the far edge of the plane makes it unnecessary to mark the dimensions of the rebate on the surfaces of the wood prior to cutting the rebate.

PLANE DETAILS

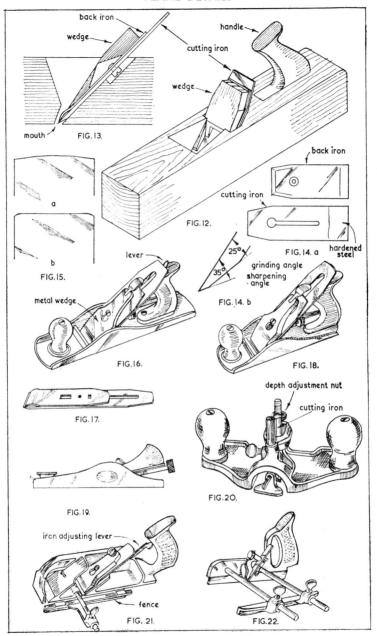

back iron

wedge

handle

cutting iron

wedge

mouth

FIG. 13.

FIG. 12.

a

b

FIG. 15.

back iron

cutting iron

FIG. 14. a

hardened steel

lever

25°

35°

grinding angle

sharpening angle

FIG. 14. b

metal wedge

FIG. 16.

FIG. 18.

FIG. 17.

depth adjustment nut

cutting iron

FIG. 20.

FIG. 19.

iron adjusting lever

fence

FIG. 21.

FIG. 22.

The plough, Fig. 22, is used for cutting grooves for panels, etc., and has a fence similar to that of the rebate plane. This is set to coincide with the position of the groove on the edge of the timber. A set of irons of varying widths is supplied with the tool, and the amount of cut can be adjusted by means of the screw seen behind the cutting iron.

SHOOTING AN EDGE WITH THE JOINTER PLANE
Note that the fingers of the left hand curl beneath the sole and act as a sort of fence.

The bullnose plane, Fig. 23, can be used for several purposes such as the cleaning up of stopped chamfers, the nearness of the cutting edge of the iron to the front of the plane making it possible to clean up the surface almost to the stopped end. The surfaces of rebates where they intersect with bevelled shoulders, too, is another instance of where this plane can be used to advantage.

The shoulder plane, Fig. 24, is used for finishing shoulders to tenons in hardwood joinery. It can also be used for cleaning up rebates, the working of mouldings on framework, and many other operations.

The spokeshave, Fig. 25, is also a type of plane, this tool being

TOOLS

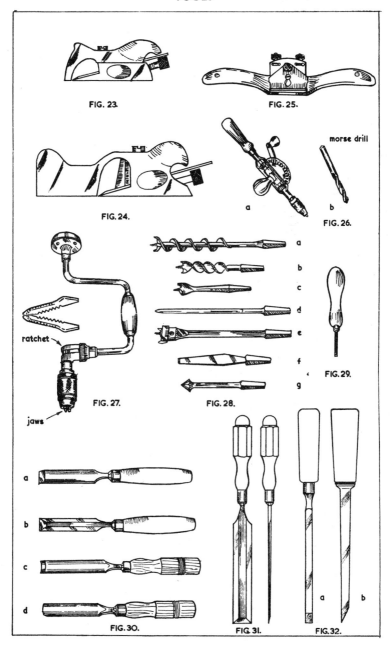

FIG. 23.

FIG. 25.

FIG. 24.

a

morse drill

b

FIG. 26.

a

b

c

d

e

f

g

ratchet

jaws

FIG. 27.

FIG. 28.

FIG. 29.

a

b

c

d

FIG. 30.

FIG. 31.

a

b

FIG. 32.

used for cleaning up curved surfaces. The spokeshave can be either flat-faced or curved. The flat type is used for cleaning up round or convex surfaces; the round-faced spokeshave is for the preparation of hollow or concave surfaces. Iron adjustment is by turning the two nuts seen above the cutting iron.

Boring tools. There are several kinds of boring tools available, the three most commonly used are the bradawl, the hand drill, and the brace.

The hand drill, Fig. 26a, is used for drilling small holes for screws, etc. Several morse drills, Fig. 26b, from $\frac{1}{16}$ in. up to $\frac{1}{4}$ in. in diameter are usually sufficient for the carpenter's normal requirements. Tungsten-tipped drills can be used in the hand drill for boring holes in brick, breeze, etc. for the insertion of plugs.

The ratchet brace, Fig. 27, is a general-purpose tool. Its uses include the boring out of mortises, recessing for shelves, boring holes for bolts, dowels, etc. and many other jobs where holes are required. It is usual to buy a brace with a ratchet, Fig. 27, so that holes in awkward corners or positions can be bored. The bit is placed in the lower end of the brace between spring-loaded jaws which grip the squared end of the bit firmly whilst the boring is taking place.

Bits. There are several kinds of bits for use with the brace some of which are illustrated in Fig. 28. At (a) is shown the twist bit which can be obtained from $\frac{1}{4}$ in. diameter up to 1 in. If larger holes are required an expansion bit (e) is used. This has two interchangeable cutters which are able to produce holes up to 3 in. in diameter. The shorter twist bit shown in (b) is often called a dowel bit, and is used for boring the holes for dowels which secure certain types of joints.

The centre bit (c) is a much cheaper bit to purchase, and can be obtained to bore holes much larger than 1 in. diameter. Many carpenters have one of these to cut $1\frac{1}{4}$ in. diameter holes specially for the purpose of fixing cylinder locks which require a hole of that size in the door stile for their fixing. They are also useful for housings and recesses, prior to finishing off with a router.

The pin or shell bit (d) is useful for boring holes quickly for screws, and the countersunk bit (g) is used for recessing the surfaces of the screw holes to allow the heads of countersunk screws to finish flush or below the surface of the wood. The screwdriver bit (f) is useful when large screws have to be turned into wood, the wide sweep of the brace making this a much easier job.

The bradawl, Fig. 29, is similar to a chisel sharpened on both sides, and is used for boring holes for small screws through thin material such as plywood by holding in the hand and applying pressure to the tool. The cutting edge of the tool should be placed across the grain of the wood so that the edge will cut through the fibres when pressed into the timber.

Chisels. The two chisels illustrated in Fig. 30 have their own particular functions.

The firmer chisel (*a*) is used for the heavier type of work such as removing surplus timber from housings and recesses, roughly shaping timber, removing large chamfers prior to planing, etc.

The bevelled edged chisel, (*b*), is for the finer work such as paring surfaces flat, preparing shoulders to tenons and dovetailed joints, etc. Sizes of these chisels can be obtained from $\frac{1}{16}$ in. up to 2 in.

The paring chisel, Fig. 31, which is longer than the bevelled edged chisel is similar in form, and is used for paring wide surfaces.

The gouges (*c* and *d*) Fig. 30, are used for curved work. The blades are curved in section. That in (*c*) is a scribing gouge and is ground and sharpened on the inside of the curve. That in (*d*), a firmer gouge is sharpened on the outside.

Fig. 32 shows two mortise chisels, and, as their name implies, are used for cutting the mortises for tenons. Their blades are much stronger than those in Figs. 30 and 31 because of the heavy type of work for which they are intended. As for all the chisels already mentioned, the handles should never be hit with a hammer. A mallet should always be used on mortise chisels and only when absolutely necessary on the other types. Mortise chisels can be obtained in widths up to $\frac{3}{4}$ in. and in some cases up to $1\frac{1}{2}$ in.

Squares. The try square, Figs. 33, 34, and 35, is used for marking out right angles such as the shoulders to tenons, and marking the positions of mortises and housings for shelves, etc. They can be obtained with blades up to 12 in. long. The square illustrated has a wooden handle with brass facing, the blade being fixed to the handle with three rivets. It is also possible to obtain all-metal try squares, the blades of which are usually marked in inches, and fractions of an inch.

It is essential that the blade is at right angles with the brass facing of the handle, and to test it obtain a piece of wood with a perfectly straight edge. Place the square across the timber with the butt against the edge, and mark a line along the outer edge of the blade on the face of the timber. Then turn the square round so that the handle is pointing in the opposite direction to see that the edge of the blade is in line with the mark on the timber. If the square is not accurate it must be corrected carefully with a file.

The mitre square, Fig. 38, is similar to the try square but this is used for marking angles of 45 degrees.

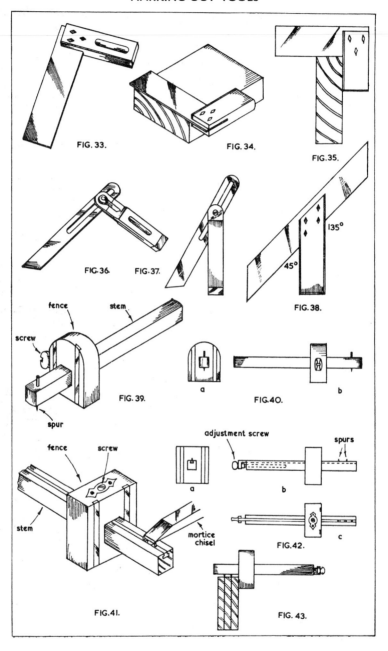

FIG. 33.

FIG. 34.

FIG. 35.

FIG. 36. FIG. 37.

135°

45°

FIG. 38.

fence   stem

screw

FIG. 39.

spur

a   b

FIG. 40.

adjustment screw   spurs

fence   screw

a   b

stem

c

mortice   FIG. 42.
chisel

FIG. 41.

FIG. 43.

The sliding bevel (Figs. 36 and 37) consists of a movable blade which can be set up to any angle. A lever, Fig. 36, or a screw, Fig. 37, is used for locking the blade in position. This tool is used for marking splayed shoulders to tenons, bevels to roof members, bevels on the edges of timber, etc.

SHARPENING CHISEL ON THE OILSTONE
The honing angle is in the region of 30 deg.

Gauges. The marking gauge, Fig. 39, is used for marking lines parallel with the timber in the direction of the grain. It has a stem about 9 in. in length on which there is an adjustable fence. A spur near one end of the stem marks a line along the timber, the distance from the edge being controlled by the fence sliding along the edge of the piece of wood. The depth of the mark can be governed by allowing the corner of the stem to be in contact with the wood being gauged. Fig. 40a and b show the end and side views of the tool.

The mortise gauge, Figs. 41, 42, and 43, is similar to the marking gauge but in this tool there are two spurs, the distance between them being adjustable as well as the position of the fence. This tool is used for marking, in one operation, the two lines required for mortises, tenons, grooves for panels, etc. The mortise chisel or plough iron is first selected, and the spurs opened to fit this dimension. Next the fence is positioned to allow the spurs to mark the lines on the timber where required. The distance between the spurs is adjusted by turning the adjusting screw at the end of the stem. The fence is secured on the stem by the screw shown in Fig. 41. Fig. 42 illustrates various views of the tool, and Fig. 43 shows it being used for marking a tenon.

Hammers. There are two hammers in common use today, the claw hammer (Fig. 44) and the Warrington or bench hammer (Fig. 46). The claw hammer is a carpenter's tool, and is seldom seen in the joiner's shop. It is most useful on the building site, and is more suitable for that type of work. It is useful for roofing and flooring work as well as formwork for concrete, and it is useful for extracting nails without resorting to pincers.

If this tool is used for pulling out nails from the surfaces of joinery work, a piece of wood or the blade of a square should be placed under the head to prevent it from damaging the surface, Fig. 45. The Warrington hammer is a joiner's tool, and usually is used for bench work. The pincers (Fig. 48) are used for extracting nails.

Mallet. The mallet, Fig. 47, should always be used in preference to a hammer when chisels and other wooden-handled tools are being used. These are made of beech and ash, and if used properly can give years of service.

Screwdrivers. Fig. 49 illustrates three kinds of screwdrivers in common use. In (a) is shown the ordinary type of screwdriver which has a blade from 2 in. to 18 in. long which is fixed to a handle of boxwood or beech. In (b) is shown a newcomer to the range of screwdrivers, and is used for turning the Philips type of screw which has a cross-shaped recess in the head. At (c) is a ratchet screwdriver, a useful tool which can operate similarly to the hand brace, and can turn screws into or extract them from the wood. The ratchet can be placed out of action and the tool used in a similar way to that in (a).

Various tools. Fig. 50 shows a pair of dividers or compasses. These are used for marking curved lines on the surfaces of timber and can also be used for scribing boards to uneven surfaces.

The glasspaper block, Fig. 53, is a piece of hardwood, usually about 4 in. long, 3 in. wide and $\frac{3}{4}$ in. thick, and is used, with glasspaper, for finishing surfaces after they have been prepared by machine or tool. The glasspaper should be torn to a convenient size and

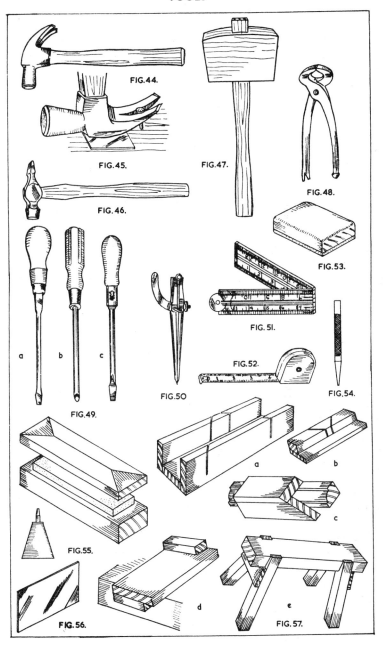

FIG.44.

FIG.45.

FIG.46.

FIG.47.

FIG.48.

FIG.53.

a b c

FIG.49.

FIG.50

FIG.51.

FIG.52.

FIG.54.

a

b

c

FIG.55.

d

e

FIG.57.

FIG.56.

placed round the block. The lower surface of the block is often covered with a piece of cork lino or similar material.

Two types of rule are shown in Figs. 51 and 52. The first, a boxwood 2 ft. rule, is usually seen in the joiner's shop, but the all-metal tape rule which can be purchased up to 10 ft. in length is a most useful tool on the building site. The hook at the end enables the craftsman to take fairly long measurements, across openings, etc., without difficulty. .

Fig. 54 shows a nail punch for knocking the heads of nails below the surface of the wood.

Fig. 55 illustrates an oilstone in a box for protection. A box similar to that shown can easily be made by the owner himself, and is well worth the trouble. A good stone should always be purchased in preference to a cheap one. A carpenter on the site would do well in obtaining one which has two grades of stone. One surface has a coarse, quick-cutting action, and the other a much finer surface for obtaining a keen edge. The joiner in the shop who usually has a grindstone at his disposal needs only a finer or medium type of stone. Fig. 55 also shows an oilcan which is necessary for supplying the oil for the stone.

Scraper. In Fig. 56 is a metal scraper used for removing all marks from the surfaces of timber, especially hardwoods, where a smoothing plane has not been able to produce the necessary surface or on veneers which are too thin to prepare by plane. The scraper is gripped by both hands with fingers in front and thumbs behind. The thumbs give a forward and downwards thrust as it is pushed over the surface of the wood with one of its longer edges in contact with the surface of the timber being prepared. The top edge of the scraper should be in advance of the lower edge, the tool making an angle of about 70 degrees to the surface of the wood.

Appliances. Fig. 57 shows several accessories necessary for the shop and site work. At (a) and (b) are two types of mitre blocks used for cutting the ends of mouldings, etc. at angles of 45 degrees. At (c) is a mitre templet which can be fixed in the bench vice with the moulding for cutting the ends of the mouldings, etc. at any angle. The templets are made to produce the required angle.

At (d) is a bench hook for resting up against the edge of the bench or for securing in the vice to enable pieces of timber to be cut across the grain easily.

Fig. 57e is a sawing stool which allows larger pieces of timber to be sawn to size, or jobs to be assembled, etc.

It is well to note that there are many other tools available, but those mentioned here should give young carpenters and joiners an idea of the tools required to carry out the usual straightforward jobs.

CHAPTER 3   COMMON JOINTS

WHEN MARKING OUT the various pieces of timber which go to form a job the person marking out must have a good knowledge of the joints which are to be used, and it is the purpose of this chapter to bring to the notice of the young joiner and carpenter a variety of joints in common use today.

Width joints. It is often necessary to make boards wider than they are when purchased, and Figs. 1–4 illustrate how this can be done. Fig. 1 is a simple butt joint. The edges of the two boards to be jointed are shot straight and square with a jointing plane, the two surfaces spread with a glue, and brought into contact. Cramps or metal dogs are used to keep the surfaces together until the glue sets. Fig. 2 is similar, but in this case a plywood tongue is inserted in the grooves made in the two edges of the boards. The edges of the boards must, of course, be shot straight and square before the glue is applied.

Fig. 3 is another butt joint with the addition of dowels. Care must be taken that the two holes for each dowel are correctly positioned, and this can be done with the aid of the try square and marking gauge. Fig. 4 is a slot-screwed joint. A countersunk screw is turned into one of the pieces to be jointed, and a slot slightly wider than the shank of the screw is cut in the other. At one end of the slot a hole is bored which will allow the head of the screw to enter. When the two pieces are brought together they should be lightly cramped, and the end of the piece containing the slot should be sharply tapped with a hammer so that the screw is forced along to the opposite end of the slot. The head of the screw will have ploughed its way through the grain of the other piece and will thus hold the joint firmly together. The joint is then knocked apart and the glue applied to both surfaces. A half-turn of the screws with the screwdriver before assembling the joint finally will ensure a close fitting joint.

Figs. 5–8 are joints used in flooring boards. The first two are rebated and tongued-and-grooved respectively. Figs. 7 and 8 are joints used when secret nailing is required, the nails being driven through the thick part of the tongue.

Corner joints. Figs. 9–16 are corner joints and can be used for the external angles of boxes, cabinets, wall panelling, etc. Fig. 9

is just a simple butt joint and would only be used on cheap work. Fig. 10 is rebated and beaded. One piece has a rebate worked on one edge and a bead is also worked on the outside edge of the rebate. Fig. 11 is a simple rebated joint.

Fig. 12 is also rebated, but instead of a bead an ovolo moulding has to be worked on the edge of the rebate. Fig. 13 is a simple mitre joint with both pieces bevelled at an angle of 45 degrees. In Fig. 14 both pieces are again mitred but this time a rebate has been introduced. In Fig. 15 a plywood tongue has been used, and Fig. 16 shows yet another method, the introduction of a tongue and a groove. Fig. 17 is a tongue-and-grooved joint with the front edges of each piece chamfered, the type of joint commonly used for matching.

Halving joints. The next three joints are angle joints and can be used where two pieces of timber running in different directions have to be secured together. They might be used for the wall plates for a roof, the framing for a portable hut, etc. Fig. 18 shows a halving joint used at a corner intersection. Half the thickness of each piece is cut away as shown and the two pieces fixed together to form a right angle. Fig. 19 is a T halving and is similar to the last joint. It is used where one piece intersects with another at any point along its length. Fig. 20 is a dovetailed halving and is obviously a much stronger joint than the last two. If properly made this joint cannot come apart due to lateral movement.

Housed joints. Fig. 21 is a housed joint and can be used for shelves in cabinets, partitions in boxes, etc. A groove is made across the width of one piece to coincide with the thickness of the other. The joint can be secured with nails and glue. Fig. 22 is also a housed joint but in this instance the groove is not allowed to pass through to the front edge of the vertical piece. This joint is used in better-class work. Fig. 23 is a stopped housed joint used in floors, where one joist relies on the support of another (see chapter on floors).

Mortice-and-tenon joints. Figs. 24–31 illustrate mortice-and-tenon joints. Fig. 24, an open mortice-and-tenon joint, is often called a bridle joint, and can be used for the intermediate members of a vertical frame. Fig. 25, the common mortice-and-tenon joint, consists of a hole in one of the pieces to be jointed (the mortice) to receive a portion of the other member (the tenon). Figs. 26–28 are three tenons, the first shows the effect of running a groove along one edge of the timber. It reduces the width of the tenon, and so allowance for this must be made when marking the width of the mortice.

A mortice-and-tenon joint at the corner of a piece of framing requires the tenon to be approximately half the width of the timber

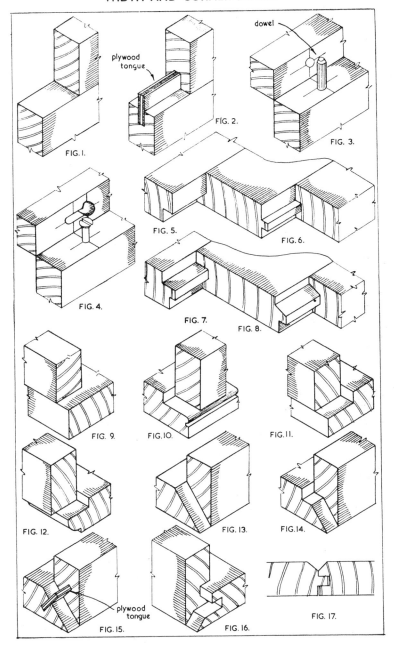

FIG. 1.

FIG. 2.

plywood tongue

FIG. 3.

dowel

FIG. 4.

FIG. 5.

FIG. 6.

FIG. 7.

FIG. 8.

FIG. 9.

FIG. 10.

FIG. 11.

FIG. 12.

FIG. 13.

FIG. 14.

FIG. 15.

plywood tongue

FIG. 16.

FIG. 17.

being used, see Figs. 27 and 28.   The other half is occupied by a haunching which fits into a groove cut above the mortice in the other piece.   Sometimes the haunching is left square as in Fig. 27 but often, in the case of good joinery, the haunching is bevelled as in Fig. 28.

Fig. 29 is the mortice-and-tenon joint used at the upper corner of the top sash in a boxed frame.   In this case the haunching is left on the piece which contains the mortice, and the portion adjacent to the tenon is housed out to receive the haunching.   This is often known as franking.   Fig. 30 is a mortice-and-tenon joint with bevelled shoulders, and is used where there may be pressure in a downwards direction on the piece which has the tenon.   The horizontal surface made by bevelling the shoulders greatly strengthens the joint.

Fig. 31 is an oblique mortice-and-tenon joint and is used in carpentry work as in roof trusses, heavy arch centres, etc.   The joint is usually secured with a bolt passing through the two pieces or a metal strap.

Angle joints.   Figs 32 and 33 are internal angle joints for mouldings and skirtings.   In cases such as these, if at all possible, the mouldings should be scribed at the intersection.   This consists of cutting the profile of one of the mouldings on the end of the other so that when they are fitted together the end of one fits exactly against the surface of the other.   Fig. 33 is a similar joint with a tongue-and-groove added.   This is used on better-class work, especially with hardwood mouldings.

Jointing table-tops and counter-tops and similar wide boards to underneath framing is carried out by means of buttons or metal plates (Figs. 35 and 37).   Fig. 34 illustrates how a button is used for securing a table-top to its framing.   A groove is made on the inside edges of the rails of the leg framing, the lip of the button is inserted into the groove and screwed to the table-top.   This will ensure the top being able to shrink or swell according to the amount of moisture in the atmosphere.

The metal plate in Fig. 37 acts in the same way.   It is first housed and screwed into the top of the rail of the framing.   The plate is then screwed to the top board, using the slot in the plate.   The slot will allow the screw to slide along its length if necessary if shrinkage or swelling takes place.   It is necessary to have some plates with slots running across their widths, as in the drawing, and some with the slots in the opposite direction.   The plate illustrated would be used across the width of the top board and the others along its length.

Length joints.   Methods of lengthening timber are important. In frames with shaped heads, for instance, the methods shown in

HALVING, MORTICE AND TENON JOINTS

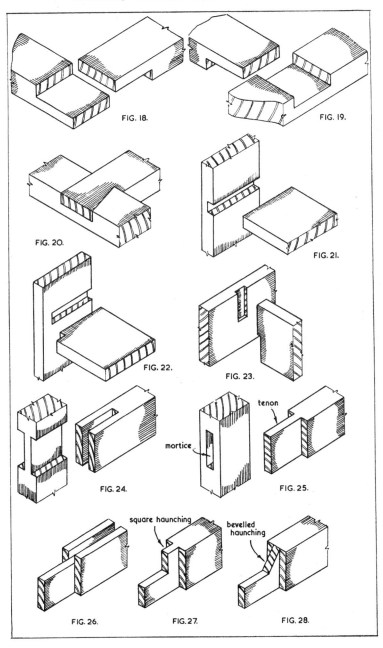

FIG. 18.

FIG. 19.

FIG. 20.

FIG. 21.

FIG. 22.

FIG. 23.

FIG. 24.

tenon

mortice

FIG. 25.

square haunching

bevelled haunching

FIG. 26.

FIG. 27.

FIG. 28.

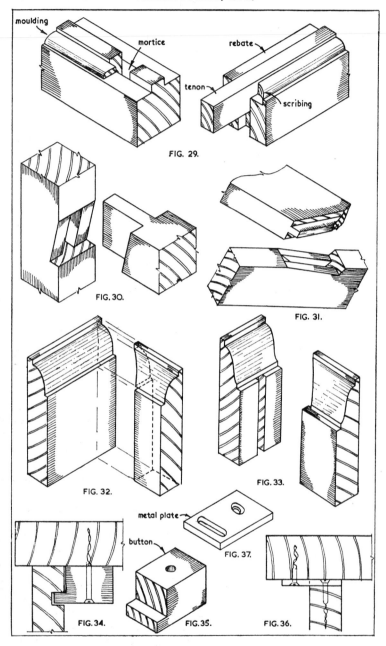

FIG. 29.

FIG. 30.

FIG. 31.

FIG. 32.

FIG. 33.

FIG. 34.

FIG. 35.

FIG. 36.

FIG. 37.

Figs. 38 or 40 could be used. Fig. 38 illustrates a hammer-headed key joint. The two pieces to be secured together are butt jointed and a hardwood key, to the shape shown, is housed into the two pieces. Two wedges are inserted under one of the heads of the key so that the joint surfaces can be brought up as close together as possible. To stop any possible twisting movement, two small tongues should be used, one on each side of the key.

Fig. 40 shows how two pieces of timber can be jointed end to end with the use of a handrail bolt. Holes are bored in the ends of each piece and small mortices are cut in the side of each so that the nuts can be inserted. The first nut is placed into one of the mortices and the bolt screwed on to it. The other end of the bolt is passed through the hole of the second piece after the round, slotted nut has been inserted in the small mortice of this piece. The nut can be made to engage the thread of the bolt by turning it with a handrail punch and the slots on the nut. A washer should also be used under each of the nuts. A dowel on each side of the bolt should also be used to stop any twisting action.

Fig. 39 shows how a wide board can be lengthened. These are called counter cramp joints and, as the name implies, are commonly used on counter-tops. Each joint consists of three pieces of batten with trenchings made in the surfaces in contact with the board to take folding wedges. Two of the battens are screwed to one part of the board and the other is screwed to the second half. The wedges are then inserted and tightened to bring the joint up tight. The other screws are driven home to secure the joint.

A study of the drawing will show that there must be a clearance gap behind and in front of the wedges. For instance there is a gap behind the wedges in the two outside battens and a gap in front of the wedges on the centre batten. If these clearances are observed and the screws shown in the drawing only are screwed home, then the folding wedges will be able to force the two ends of the top together. Then, and only then, are the other screws used.

Figs. 41 to 43 show two other methods used for lengthening timber and are used in constructional work. In Fig. 41 several pieces of wood are placed together in, say, three layers, with the joints staggered, and then they are nailed or bolted together.

Fig. 42 illustrates how fairly heavy timbers are lengthened. The ends are butt jointed and wooden fish-plates are bolted across the joint as shown. The same principle applies in Fig. 43 but in this case metal fish-plates have been used in place of the timber.

Dovetails. Figs. 44 to 48 illustrate various types of dovetail joints. The simple dovetail joint in Fig. 44 is used for securing the

COMMON JOINTS

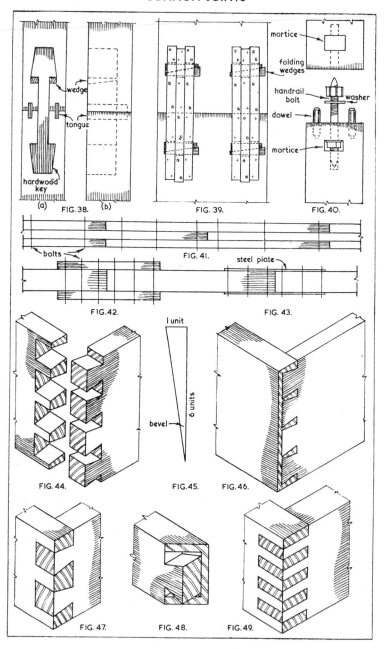

wedge
tongue
hardwood key
(a) FIG. 38. (b)

mortice
folding wedges
handrail bolt — washer
dowel
mortice
FIG. 40.

FIG. 39.

bolts
FIG. 41.

steel plate
FIG. 42.
FIG. 43.

FIG. 44.

1 unit
6 units
bevel
FIG. 45.

FIG. 46.

FIG. 47.
FIG. 48.
FIG. 49.

DOVETAILS

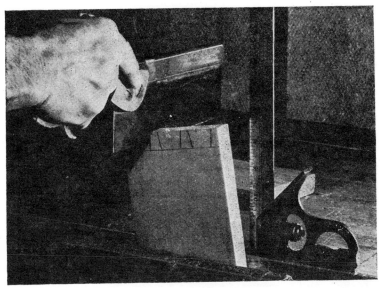

SAWING THE DOVETAILS

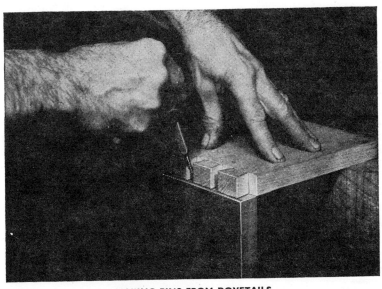

MAKING PINS FROM DOVETAILS

PINS WITH WASTE PARTLY SAWN AWAY

sides of boxes, carcasing for counters and similar fittings, etc. There are several bevels for dovetails in common use, and one which is considered adequate by many people is a bevel of 6 in 1 (Fig. 45). If a right angle is constructed with one arm of the angle 6 units in length and the other 1 unit long the diagonal drawn to the ends of the two arms will give the bevel.

When dovetail joints are to be used but it is not desirable for the joint to be seen on the face of the work, a lapped-dovetailed joint can be used (Fig. 46). This joint is often used for the front of drawers, tops of cabinets, etc. and, as the drawing shows, the front portion of the work is allowed to overlap the front surface of the dovetails and so completely hides them from view seen from the front position.

Dovetails are used in carpentry work, such as in the trimming round floor-openings, and these are usually made so that the pins are the same width as the dovetails for strength reasons (Fig. 47). Fig. 48 shows one portion of a secret dovetailed joint, and shows that the dovetailed portion of the joint is completely enclosed in the mitred part.

Fig. 49 is similar to the dovetailed joint and is called a comb joint and only used in cheaper work. The interlocking portions are square.

CHAPTER 4   FIXING DEVICES

THE COMMONEST OF fixing devices is probably the nail.

Nails.   There are many types of nails available, and each is more suited to one particular job than for other purposes.   A small selection of nails is shown in Fig. 1.

Fig. 1a shows a french wire nail.   It is fairly stout in comparison with its length to withstand heavy hammering without bending.   It is used mainly for carpentry work such as nailing concrete formwork in place, roof work, temporary work such as centers for arches and shoring, and many other jobs of a similar nature.   It is sometimes advisable to bore a hole to receive a wire nail to prevent the timber from splitting, especially in hardwoods.

At (b) is an oval brad.   As the name implies, these nails are elliptical in section.   They have less tendency to split the timber if they are nailed into the wood with the widest dimension of the section in the same direction as the grain of the timber.   These nails are used for carpentry as well as joinery work, and in the latter the heads are easily punched below the surface of the work without much damage.   They are used for jobs such as the fixing of mouldings, securing joints in boxes and casings, fixing matchboarding, etc.

At (c) is a floor brad used mainly for fixing floor-boards to the joists.   This has a blunt end which tears away the wood fibres as it passes through the board, and so avoids splitting.   It also has a wedging action, and securely fixes the boards down on to the joists.

At (d) is shown a nail similar to the wire nail but with a much smaller head.   It is rounded as shown.   This is the lost-head nail, and is widely used in joinery work where the head has to be punched below the surface without making too large a hole.

The panel pin, (e), too, is similar to these last-mentioned nails but is much smaller in section.   These can be used for work such as fixing plywood, hardboard, mouldings, and other light work where strength is not so important.   If strength is required panel pins can be used in conjunction with one of the glues which are mentioned later in this chapter.

Clout nails, (f), which are usually galvanised, are used for fixing roofing felt.   They have wide heads to prevent the material they are fixing from tearing loose.

FIXING DEVICES

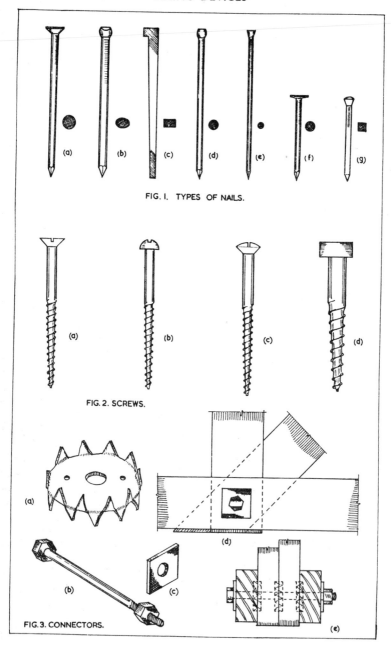

FIG. 1. TYPES OF NAILS.

FIG. 2. SCREWS.

FIG. 3. CONNECTORS.

The small nail at (*g*) is used for fixing hardboard or thin plywood, and is made of copper. The heads are so shaped so that they can be knocked below the surface without having to be punched below.

Screws. There are also many kinds of wood screws available. Four of them are shown in Fig. 2. The first is the commonest, and is called a countersunk screw (*a*). Screws, obviously, are a much stronger fixing than nails, and where great strength is required should be used in preference to them.

It must be remembered that where screws are used, pilot holes must first be bored either with a brace and bit or a wheelbrace and morse drill. If countersunk screws are to be used the top of the pilot hole should be countersunk to receive the head of the screw so that it can either finish flush or below the surface of the wood.

Often, where this type of screw is used, the screw head is turned well below the surface, and the hole filled with putty or a specially prepared filler, or small round pellets of wood are made to fit the hole. The pellets are glued and tapped into the hole, and when the glue has set the top of the pellet is cleaned off with a smoothing plane flush with the surface of the wood.

Round-head screws (*b*) are used in positions where the screws are exposed to view and have to be removed periodically, and also on thin metal fittings where it is not practicable to countersink the surface of the metal such as shelf brackets, rim locks, etc.

At (*c*) is shown a raised-head which is countersunk as well as having a rounded head. These are used on good quality work on parts which have to be removed if necessary, such as the glazing beads to a glass-fronted counter, traps to heating casings, in exposed hinges such as counter flaps, etc.

At (*d*) is a coach screw used for holding heavy timbers, fixing metal plates, etc.

Timber connectors. Where several thicknesses of timber have to be bolted together, such as in the manufacture of the modern type of roof truss, the bolted joints are greatly strengthened by introducing between each pair of timbers what is called a timber connector. One of these is seen in Fig. 3*a*. There are different types, but that illustrated is the most commonly used. It is about 2 in. in diameter and has teeth similar to saw teeth projecting on each side. In the centre is a hole through which the bolt (*b*) passes.

At (*e*) is shown a cross-section through the joint involving four pieces of timber, and it can be seen that three connectors would be required for this joint. The teeth of the connectors are forced into the timber whilst tightening the nut of the bolt with a spanner. To prevent the nut and head of the bolt from being forced into the timbers large square washers (c) should be used.   At (d) is seen an elevation of the joint.

Wooden plugs.   When fixing joinery items to a wall, wooden plugs shaped similarly to Fig. 4a can be knocked into raked out joints in the brickwork, and the item to be fixed nailed to the plugs. A handy tool suitable for shaping the plugs is a carpenter's axe.   If shaped as shown the plugs have a wedging action, and are securely fixed if the work is carried out in a workmanlike manner.   The joints in the brickwork can be raked out with a plugging tool as shown in Fig. 4b.

If grounds for panelling are to be fixed to a brick wall, several joints should be raked out at intervals across the wall and in varying heights between the floor level and top of the ground work (d). The plugs when fixed and ready to receive the grounds must have their front surfaces perfectly plumb and all in line with each other across the width of the room.   To do this the plugs at the extreme ends must first be made plumb by trimming off their ends so that when a plumb rule is placed on their front surfaces the plumb bob will be dead centre, as seen in (c).

When the plugs at each end of the wall have been cut and made plumb, a string should be stretched from one end to the other as seen in Fig. 4e so that the correct positions for cutting the intermediate plugs can be marked.   When these have been trimmed off to the lines the wooden grounds can be fixed to the plugs by nailing.

The most efficient way of fixing items to solid walls with screws is to first bore a hole into the wall with the special drill or plugging tool, and insert a fibre plug, or, if there is a danger of dampness, one of soft metal, see Fig. 5a and b.   The item is fixed by screwing into the plug.   The special tool for this work, Fig. 5c, should be used to make the hole.   Plugs and tool bits to suit the size screw should always be used.

The method of making the hole is shown in Fig. 5f.   The point of the tool should be placed on the wall in the position where the hole is required.   The hole is made by tapping the tool with a hammer and rotating it as the hammering is taking place.   Care should be taken to hold the tool at right angles to the face of the wall.   When the hole has been cut to the correct depth the plug should be pressed gently into the hole when it is then ready to receive the screw.

Many craftsmen prefer a durium-tipped masonry drill (e) for making the hole for the plug.   If kept in good condition, this will cut through brick, tile, marble, cement, and other materials.   Different sizes can be obtained for different sizes of screws.

Toggles.   When fixing items to hollow walls and partitions it is

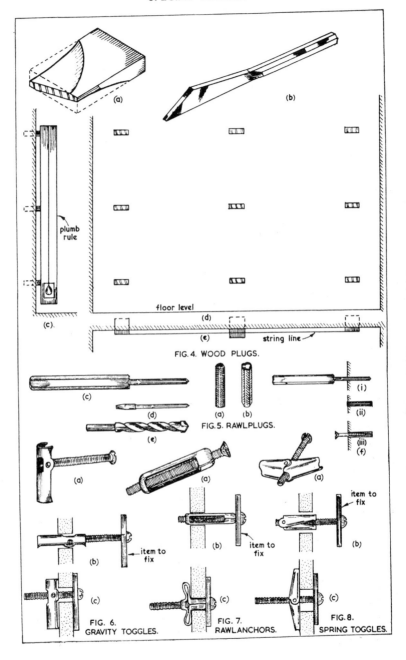

(a)

(b)

plumb rule

(c).

floor level

(d)

(e)   string line

FIG. 4. WOOD PLUGS.

(c)

(d)

(a)   (b)

(i)

(ii)

(iii)
(f)

(e)

FIG. 5. RAWLPLUGS.

(a)

(a)

(a)

(b)

(b)   item to fix

item to fix

item to fix

(b)

(c)

(c)

(c)

FIG. 6.
GRAVITY TOGGLES.

FIG. 7.
RAWLANCHORS.

FIG. 8.
SPRING TOGGLES.

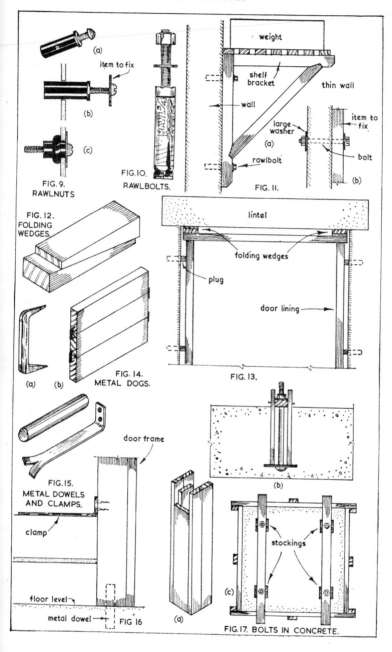

FIG. 9.
RAWLNUTS

item to fix

(a)
(b)
(c)

FIG. 10.
RAWLBOLTS.

weight

shelf bracket

thin wall

wall

large washer

item to fix

bolt

rawlbolt

(a)

(b)

FIG. 11.

FIG. 12.
FOLDING WEDGES.

lintel

folding wedges

plug

door lining

(a)   (b)

FIG. 14.
METAL DOGS.

FIG. 13.

door frame

FIG. 15.
METAL DOWELS AND CLAMPS.

clamp

floor level

metal dowel

FIG 16

(a)

stockings

(b)

(c)

FIG. 17. BOLTS IN CONCRETE.

necessary, if a good secure job is required, to use the correct type of fixing. The Rawlplug Co. have produced some useful devices called toggles which are ideal for this type of work. At Fig. 6a is a gravity toggle. This is useful for fixing items to a material of $\frac{1}{2}$ in. or less in thickness. A hole should first be bored through the material to which the item is to be fixed. It should be slightly larger than the body of the toggle. The screw should be removed from the body and passed through the hole in the item to be fixed. The body should then be engaged by the screw. The body of the toggle is passed through the hole in the partition as in Fig. 6b when the body will fall into position as in (c). The screw can then be turned to secure the item.

Rawlanchors, Fig. 7a, are a similar type of fixing device and can be used for fixing items to hollow bricks or tiles, hollow partitions, walls filled with foamed slag wood wool, and other loose materials. The rawlanchor is secured to the item to be fixed similarly to the gravity toggle, and passed through a hole which has been bored in the hollow wall or partition. With the item pressed up against the partition in the position where it is required, the screw should be turned with a screwdriver, so causing the back of the anchor to bend and flatten up against the inside surface of the partition as seen in (c).

Spring toggles, Fig. 8a are similar to the gravity toggle. In this the toggle is spring-loaded, and is suitable in positions where the gravity toggle would not work such as in horizontal positions, i.e. ceilings. A hole about $\frac{1}{2}$ in. diameter is first drilled into the horizontal, or any other thin surface for that matter. The screw should be removed from the toggle and be inserted through the item to be fixed. The toggle is engaged once again by the screw. The toggle parts should be pressed together over the screw and passed through the hole in the thin material (b). As it clears the material the halves will spring open, when the screw can be turned to secure the article to be fixed.

Another similar type of fixing is the Rawlnut (Fig. 9a). It consists of a rubber bush into which is bonded a threaded metal insert. When the screw is turned the metal insert causes the rubber to be drawn towards the partition material to which the item is to be fixed, (c). The rawlnut provides an ideal fixing for electrical goods, etc., and forms an airtight, waterproof, and insulated joint. It also moulds itself to non-flat surfaces. It is fixed similarly to the toggle bolts (b).

When heavy fixtures have to be secured to walls, such as porch roofs over entrance doors, or shelf brackets to support heavy loads,

Fig. 11, a heavier type of fixing device is required. This type of work can be carried out satisfactorily by using Rawlbolts, Fig. 10. The exterior portion of the rawlbolt consists of a tubular metal shell which is divided into four segments, the upper portions being held together with a steel ferrule. The specially shaped head at the lower end of the bolt forces open the four segments when the nut is turned with a spanner, causing them to securely grip the sides of the hole in the brickwork or other material to which the bracket is being fixed.

The first step is to bore the holes in the brickwork to coincide with the holes in the bracket. The rawlbolts are pushed into the holes, the bracket placed in position over the bolts, washers and nuts placed on the bolts, and the nuts tightened with a spanner. A fixture can be made to a thin wall in the manner shown in Fig. 11b. A bolt long enough to pass through the wall should be used and a large washer used at the rear behind the nut. This, of course, can only be used if the sight of the fixing at the rear is not objected to.

Folding wedges. These, Fig. 12, play a big part in the life of a carpenter and joiner. They are used on many occasions such as adjusting shoring and supports to concrete formwork, fixing door and window frames, supporting the ends of herring-bone strutting in between floor-joists, and many others.

Fig. 13 shows one method of assisting in the fixing of a door lining. Wooden plugs are built into the wall as the opening is being constructed, and when the linings are ready to be fixed the plugs are trimmed off so that their ends are perfectly plumb and the distance between the opposite plugs is equal to the outside width of the linings. The linings are then placed between the plugs and plumbed. Nails are driven through the linings into the plugs and folding wedges inserted immediately above the jambs of the linings and made tight. Nails should also be used for fixing the wedges to prevent them from becoming loose with the constant closing of the door. Occasionally linings are fixed similarly to external frames.

External door frames, Fig. 16, are usually placed in position when the brickwork is being built, and the fixings used are built in as the work proceeds. When cills to the frames are omitted, metal dowels can be fixed to the lower ends of the posts of the frame. The lower ends of the dowels fit into holes bored into the floor-boards; or, if the floor is concrete, the dowels fit into holes cut into the concrete and later grouted in with cement and sand slurry. As the brickwork proceeds the bricklayer will fix metal clamps on the side of the frame, see Figs. 15 and 16, and these are built into the joints of the brickwork, making, in all, a secure job. Sometimes the ends of the head of the

FIXING DEVICES

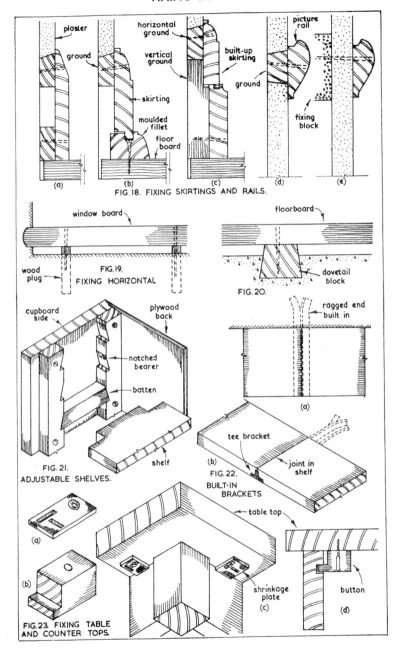

FIG. 18. FIXING SKIRTINGS AND RAILS.

(a) plaster, ground

(b) skirting, moulded fillet, floor board

(c) horizontal ground, vertical ground, built-up skirting, ground

(d) picture rail, fixing block

(e)

FIG. 19. FIXING HORIZONTAL
window board, wood plug

FIG. 20.
floorboard, dovetail block

FIG. 21. ADJUSTABLE SHELVES.
cupboard side, plywood back, notched bearer, batten, shelf

FIG. 22. BUILT-IN BRACKETS
(a) ragged end built in
(b) tee bracket, joint in shelf

FIG. 23. FIXING TABLE AND COUNTER TOPS.
(a)
(b)
(c) table top, shrinkage plate
(d) button

door frame are allowed to protrude beyond the posts, and these ends are also built into the brickwork.

Sometimes it is the job of the carpenter to provide the formwork for the concrete bed of a piece of machinery, see Fig. 17. Usually this consists of a four-sided box with the sides equal in depth to the thickness of concrete required. He may also be asked to provide a means of keeping in place the bolts which are to hold the machine down. It must be remembered that, after the machine bed has been completed, it should be possible to adjust the bolts slightly so that the holes in the machine base will fit over the bolts. To do this, what are known as stockings must be made, one for each holding-down bolt for the machine, Fig. 17a. Each one consists of four pieces of thin wood, say $\frac{3}{8}$ in. thick, to make a box of, say 3 in. square. The sides of the box should be sawn down their centres, and the two pieces held together with thin battens lightly tacked over the joints. The stockings should be held in their correct positions in the box by battens nailed across the sides of the box, Fig. 17c. The bolts are passed through the centres of the stockings with large square plates at their bases, and the tops of the bolts pass through holes in the battens and secured with their nuts. When the box has been filled, and the concrete set, the formwork can be dismantled.

The stockings are easily removed by levering off the battens on their inside surfaces and pulling out the pieces. It will be found that the bolts can be moved if necessary to coincide with the holes in the machine base. When the machine has been fixed the bolt holes in the concrete are grouted in with sand and cement. Fig. 17b shows a vertical section through the concrete with a stocking still in place.

Fixing skirtings and picture rails. Fig. 18 shows methods of fixing skirting boards and picture rails. At (a) a skirting has been fixed to grounds which have been secured to the brickwork by wood or Rawlplugs. A much better way of fixing a skirting board which prevents an unsightly gap at the bottom from appearing, due to shrinkage of the floor-joists below is seen in Fig. 18b. This method shows that the skirting is fixed at the top only. It has a tongue worked on its lower edge, and this fits into a groove in a moulded fillet which is screwed to the flooring. The shrinkage which is bound to occur is taken up in the tongue and groove, and so unsightly gaps are avoided.

When built-up skirtings (c) are to be fixed it may be more convenient go fix them in the manner shown in the drawing, especially when the back surfaces of the skirting are not all in the same plane.

CHAPTER 5 CRAFT GEOMETRY

TO THE CARPENTER and the joiner, craft geometry and a good knowledge of constructional details are as important as one another. Although the geometry contained in this chapter is of an elementary nature, it is considered sufficient, together with practical geometry in other chapters, to meet the requirements of the basic craft certificates of the various examination boards.

Bisecting. To bisect a line a–b (Fig. 1). Draw the line a–b, and with compass point in a and b in turn describe arcs above and below the line to intersect as shown in the drawing. Draw a straight line to pass through the intersections and a–b in c.

To bisect an arc of a circle, a–b (Fig. 2), follow the same directions as for Fig. 1.

To construct a right angle (90 degrees) at one end of the line a–b (Fig. 3). With compass in a (or b) and open any distance describe the semi-circle d–e, e being on the extended line a–b. With compass open a little more and with centres d and e in turn describe two arcs to intersect in f. Draw a line from a to pass through f. b–a–f is a right angle.

To construct a perpendicular line a–b from point c (Fig. 4), follow the same directions as for Fig. 3—this time using centre c to draw the semi-circle d–e.

An angle of less than 90 degrees is an acute angle (Fig. 5), and an angle of more than 90 degrees is an obtuse angle (Fig. 6).

Scales. It is sometimes necessary to make a drawing much smaller than the object it portrays. This is called drawing to scale. When the drawing is completed it must look exactly the same as the object in miniature. For this purpose a scale rule is used which contains scales of up to 3 in. to 1 ft. An example of this is shown in Fig. 8. This is one end of a scale rule and shows the scale of 1 in. to 1 ft. Each inch represents 1 ft. The first foot is divided up into twelve to represent inches. The drawing also shows how the lengths 2 ft. 6 ins. and 1 ft. 2 in. are measured from the scale.

To construct one's own scales it is first necessary to divide any given line into any number of equal parts (Fig. 7). To divide line a–b into, say, seven equal parts, first draw line a–b and from a draw line a–c at any acute angle. Mark off seven equal spaces with a rule or divider and number them 1–7. Join 7 to b and from all the other points on a–c draw lines, parallel to 7–b, to cut a–b. Line a–b is then divided into seven equal parts.

To construct a plain scale of $\frac{2}{3}$ in. to 1 ft., and 3 ft. in length, Fig. 9, draw a rectangle 2 in. long and any depth, say $\frac{1}{2}$ in., and divide it into three equal parts, each to represent 1 ft. Using the method given in Fig. 7, divide the first section into twelve equal parts to represent inches. The dimension 1 ft. 8 in. is shown on the drawing.

To construct a diagonal scale of 1 in. to 1 yd. and 2 yd. in length draw a rectangle 2 in. in length and say, 1 in. in depth and divide it into two equal parts. Divide the depth of the rectangle into twelve equal parts as in Fig. 7, and draw horizontal lines through the length of the rectangle from all these points. Divide the second yard into three equal parts to represent feet and draw a diagonal in each of the three sections as shown. The dimensions to be taken from the scale must be measured along the horizontal lines. For instance the dimension shown 1 yd. 1 ft. 5 in. is measured along the horizontal line number 5 (the number of inches in the dimension). 2 ft. 9 in. would be measured along the horizontal line numbered 9, and so on.

Quadrilaterals. Quadrilaterals are plain figures which have four straight sides. There are six of them. The first is a square (Fig. 11) which has all its sides and all its angles equal. There is a total of 360 degrees in the four corners of a quadrilateral, therefore the square has four angles of 90 degrees. Its diagonals also intersect at 90 degrees.

The rectangle (Fig. 12) has all its angles equal and opposite sides equal in length.

The rhombus (Fig. 13) has all its sides equal in length and opposite angles equal.

The rhomboid (Fig. 14) has its opposite sides and opposite angles equal.

The trapezoid (Fig. 15) has only two of its sides parallel, and the trapezium (Fig. 16) has no two sides parallel.

Triangles. There are four kinds of triangles; the right-angled triangle which contains one angle of 90 degrees; the equilateral triangle which has all its sides and all its angles equal; the isosceles triangle, which has two sides and two angles equal; and the scalene triangle which has all its sides and all its angles unequal (Figs. 17–20). There are 180 degrees in the three angles of any triangle.

Polygons. Polygons are figures with more than four straight sides. If all the sides are the same length they are called regular polygons; if they are unequal the figures are called irregular polygons. Each have their own particular names depending on the number of sides they have.

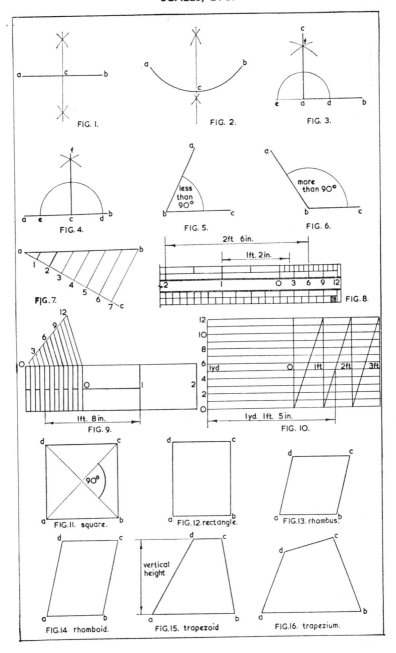

FIG. I.

FIG. 2.

FIG. 3.

FIG. 4.

FIG. 5.

FIG. 6.

FIG. 7.

FIG. 8.

FIG. 9.

FIG. 10.

FIG. 11. square.

FIG. 12. rectangle.

FIG. 13. rhombus.

FIG. 14 rhomboid.

FIG. 15. trapezoid

FIG. 16. trapezium.

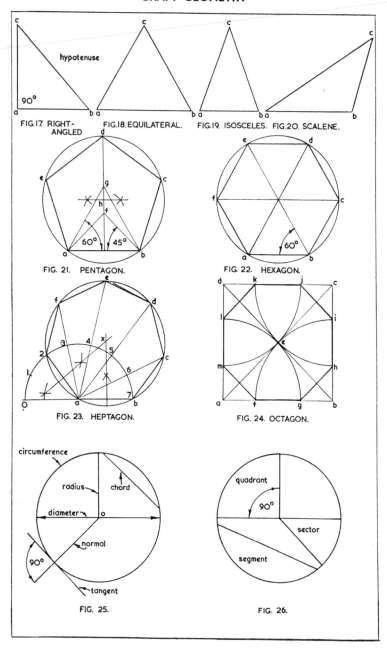

FIG.17 RIGHT-ANGLED

FIG.18. EQUILATERAL.

FIG.19. ISOSCELES.

FIG.20. SCALENE.

FIG. 21. PENTAGON.

FIG. 22. HEXAGON.

FIG. 23. HEPTAGON.

FIG. 24. OCTAGON.

FIG. 25.

FIG. 26.

Fig. 21 is a five-sided regular polygon called a pentagon. To construct a pentagon. Draw one side a–b, and on it draw two triangles the first with sides inclined at 45 degrees and the other with sides at 60 degrees. Bisect f–g to give point h which is the centre of the circle the pentagon will just fit into. With centre h and radius h–a describe a circle and step off the distance a–b five times round the circumference.

Fig. 22 is a hexagon, a six-sided figure. To construct a hexagon with sides a–b, open the compasses a–b and describe a circle and step off the distance a–b round the circumference of the circle six times.

To construct the heptagon (Fig. 23) draw the base-line a–b any required length, and with centre a and radius a–b describe the semi-circle to give point o on the extended line a–b. Divide the semi-circle into seven equal parts and number the points o–7. Draw the second side of the polygon from a to point number 2. Bisect a–b and a–2 to give x which is the centre of the circle into which the polygon will just fit. Draw the circle and from a draw lines through points 3, 4, 5, and 6 to give the positions of the other corners of the polygon.

It is important to note that a polygon of any number of sides can be constructed by using the last method. The semi-circle must be divided into a number of equal parts depending on the number of sides the polygon is to have. And the second side of the polygon must always be drawn from a to point number 2.

To construct an eight-sided polygon, an octagon (Fig. 24), in a given square, first draw the square and the diagonals to the square which intersect in e. With compass point in the four corners of the square, in turn, and with radius a–e, describe arcs to cut the sides of the square in f, g, h, etc. to give the corners of the octagon.

The circle. Figs. 25 and 26 illustrate the various characteristics of the circle. The quadrant, sector, and segment are all areas of the circle. The quadrant is exactly a quarter of a circle—the two radii forming an angle of 90 degrees. The sector can be more or less than a quadrant and is bounded by two radii and part of the circumference, and the segment is a portion of a circle bounded by a chord and part of the circumference. A semi-circle is half of a circle.

To draw a circle through any three points (Fig. 27) bisect the distances 1–2 and 2–3. The bisecting lines meet in o which is the centre of the circle.

To find the centre of a circle when two chords are given (Fig. 28) bisect the two chords 1–2 and 3–4, the bisecting lines meeting at the centre of the required circle.

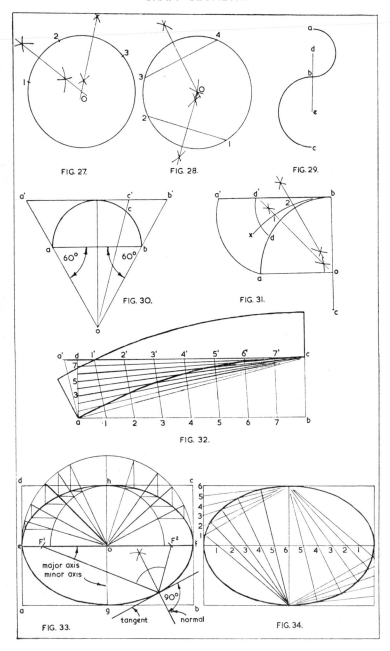

FIG. 27.

FIG. 28.

FIG. 29.

FIG. 30.

FIG. 31.

FIG. 32.

FIG. 33.

FIG. 34.

Contacting arcs must have common normals if no irregularities are to occur throughout the length of a multiple curved line. The curved line a–b–c (Fig. 29) has been drawn with two centres, d and e. These two centres must be on a straight line which passes through b which is the point of contact between the two arcs.

To find the length of the circumference of a circle (Fig. 30) draw the semi-circle a–b, and draw a 60-degree line through each end of the diameter to meet in o. If the 60-degree lines are extended upwards to meet a horizontal line drawn across the top of the semi-circle and just touching it, the length a′–b′ is equal to half the circumference of a circle with a diameter equal to a–b. If point c on the circumference is required to be transferred to a′–b′, draw a line from o through c to give c′ on a′–b′.

A more accurate method for finding the length of the circumference of a circle is shown in Fig. 31. Let a–o be the radius of the circle. Make o–c equal half o–a. With centre c and radius c–b describe the arc b–x. Bisect a–b to give point 1 on b–x. With centre 1 and radius 1–a describe the arc a–a′ to meet the horizontal line brought out from b. a′–b is equal to the arc a–b.

To transfer point d on to the line a′–b, bisect d–b to give point 2 on b–x. With centre 2 and radius 2–d describe the arc d–d′ to give the position of d on a′–b.

It is sometimes necessary to draw a portion of the circumference of a circle without the use of compasses (Fig. 32). Draw the rectangle a–b–c–d and the diagonal a–c. Make angle c–a–a′ a right angle. Divide a–b into any number of equal parts and a′–c into the same number of equal parts. Join 1–1′, 2–2′, 3–3′, etc. with straight lines. Divide a–d into the same number of equal parts as a–b and draw straight lines from these points to c. The line 1–1′ will intersect with the diagonal 1–c to give a point on the curve, 2–2′ will intersect with 2–c, and so on to give a series of points on the arc through which a freehand curve should be drawn. The curve a–c is part of a circle. The curve above a–c is a parallel curve to a–c and is obtained by opening the compasses any required distance and making a series of arcs by placing the point in several positions on a–c and drawing another freehand curve to just touch the arcs.

Ellipses. Two methods for constructing ellipses are given in Figs. 33 and 34. The first (Fig. 33) is called the auxiliary circles method. Construct the rectangle a–b–c–d and draw in major axis e–f and minor axis g–h. With centre o draw two circles, the first with radius o–e and the second with radius o–h. Draw a series of straight lines radiating from o to pass through the two circles. Horizontal lines should be drawn from where the radiating lines pass through the small circle to intersect with vertical lines drawn from the points where the lines pass through the large circle. These intersections are points on the curve of the ellipse.

The second method (Fig. 34) is called the intersecting lines method. Draw the rectangle and place in the major and minor axes. Divide each portion of the rectangle and major axis up into the same number of equal parts as shown in the drawing. Lines 1 and 1, 2 and 2, etc. intersect to give points on the curve of the ellipse.

To draw a normal and a tangent to an ellipse it is first necessary to find the focal points. To do this open the compasses half the length of the major axis and place the point in the top or bottom of the minor axis and make a mark near each end of the major axis. These are focal points (see Fig. 33). Select the point on the curve through which the normal and tangent must pass and draw straight lines from this point to the two focal points. Bisect the angle set up by these two lines. The bisecting line is the normal required, and a line at right angles to the normal to pass through the same point on the curve is the tangent required.

Inscribed circles. Many tracery panel designs rely on inscribed circles in regular and irregular figures for their basic patterns. Figs. 35 to 42 are illustrations to show how these circles are inscribed in various figures. Figs. 35 to 40 need little explanation as it is fairly easy to follow the drawings. Figs. 41 and 42 are, however, more difficult. Construct the quadrant in Fig. 41, and draw a series of lines, all, say $\frac{1}{4}$ in. apart parallel to the base line. Draw another series of lines, all the same distance apart as before parallel with the curved side of the figure. Lines 1 and 1, 2 and 2, 3, and 3 intersect to give points through which a curved line should be drawn. This curved line bisects the lower left-hand corner of the figure. Bisect, with the compasses, the 90-degree angle of the figure, the bisecting line meeting the curved bisector in o. This is the centre of the circle which will just fit into the quadrant.

Construct the figure shown in Fig. 42 and draw a vertical centre line through the figure. The problem is to inscribe the circle near the top of the drawing. This is done by drawing two series of lines as before, the first series parallel to the large curve which outlines the figure to the left, and the second series parallel to the small curve to the left of the centre line. The intersections obtained should be used for drawing the curved line which intersects the centre line in o. This is the centre of the required circle.

Mouldings. Fig. 43 shows the various Roman mouldings, and, as can be seen, all are set out with the aid of the compasses. The

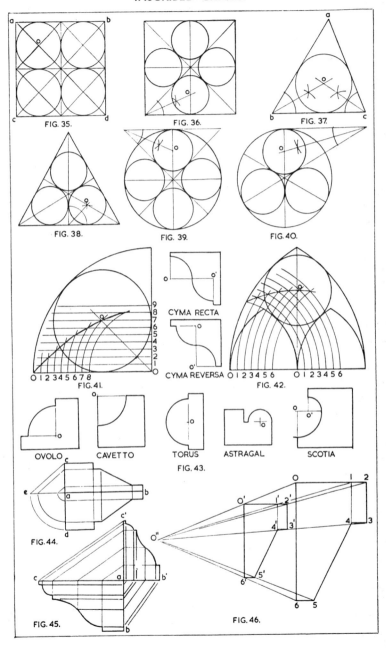

FIG. 35.

FIG. 36.

FIG. 37.

FIG. 38.

FIG. 39.

FIG. 40.

CYMA RECTA

CYMA REVERSA

FIG. 41.

FIG. 42.

OVOLO

CAVETTO

TORUS

ASTRAGAL

SCOTIA

FIG. 43.

FIG. 44.

FIG. 45.

FIG. 46.

CRAFT GEOMETRY

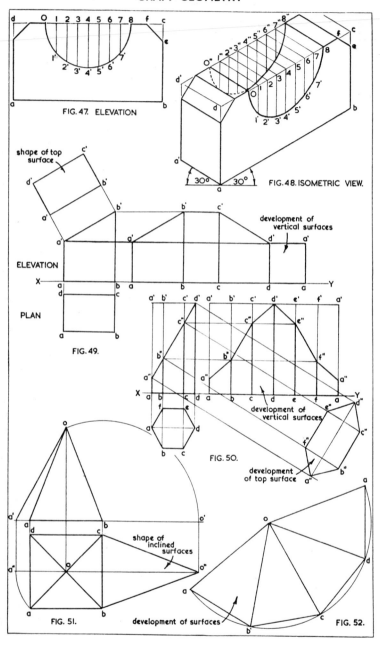

FIG. 47. ELEVATION

FIG. 48. ISOMETRIC VIEW.

shape of top surface

development of vertical surfaces

ELEVATION

PLAN

FIG. 49.

development of vertical surfaces

FIG. 50.

development of top surface

shape of inclined surfaces

FIG. 51.

development of surfaces

FIG. 52.

torus and astragal mouldings are similar. The name torus is usually applied to large mouldings with a semi-circular section, and the astragal to smaller mouldings such as the bead.

The next three drawings illustrate how mouldings are enlarged or reduced in size. In Fig. 44 the moulding a–b–c is given, and a similar moulding is required with the same width but with the thickness enlarged to a–d.

In Fig. 45, the moulding a–b–c is given, and in this case it has been necessary to reduce both the width and thickness as seen in moulding a–b′–c′. If reference is made back to Fig. 7 it will be seen that a–c′ has been reduced from a–c, and the various sections on a–c′ have also been reduced so that they are proportionately the same as an a–c. a–b′ has been reduced from a–b in a similar way.

In Fig. 46 the thickness of the moulding has been reduced in the same proportion as the width. In a problem such as this one moulding would be given and the width of the second would also be given. To construct the drawing, first draw the given moulding and a short distance away the width of the second moulding, o′–6′ should be drawn parallel to the o–6 line of the given moulding. Draw a line through points o and o′ and another to pass through 6 and 6′ to meet in o″. Draw lines from points 2, 3, 4, and 5 to meet in o″. From o′ draw a line, parallel to o–2 until it meets the line 2–o″. This is point 2′ on the required moulding. Draw another line from 2′, parallel to 2–3 until it meets line 3–o″ to give point 3′ on the required moulding, and so on.

It may be found difficult to establish the exact position of point 4′. If this is so a vertical line from 4 to give point 1 on the given moulding should be drawn. Draw a line from 1 to o″ to give 1′ on the required moulding. A vertical line from 1′ downwards to meet a horizontal line from point 3′ will give the exact position of point 4′ on the required moulding.

Isometric projection. Pictorial drawings are an important feature of architectural and building detailed drawings, and the method commonly used for isometric drawings is shown in Fig. 48. Fig. 47 shows the elevation of a block of wood, a–b in length, b–c in height, and with a semi-circular disc cut away from the top edge. The corners at c and d, too, are cut off. It is required to construct an isometric view of the block. The first step will be to divide o–8 into any number of equal parts and to draw vertical lines downwards to give 1′, 2′, 3′, etc. on the curved portion of the block.

In isometric views of objects the vertical lines remain vertical and horizontal lines are drawn at 30 degrees. Using these rules draw an isometric view of the block of wood before it is shaped. a–b and c–d are both drawn at 30 degrees, and b–c and a–d are both vertical. The thickness of the block, a–a′ is any convenient dimension. The corners can be cut off by measuring f–c and c–e on the elevation and transferring these dimensions to the appropriate lines in the isometric drawing. The second corner can be treated in the same way.

The positions of o and 8 can be marked on the isometric drawing by measuring the distances d–o and d–8 and transferring these dimensions to the isometric drawing. The distance o–8 can be divided up into the same number of spaces as in the elevation and vertical lines dropped vertically downwards as in the drawing. The vertical lines in the isometric drawing have to be exactly the same length as in the elevation, and when points 1′, 2′, 3′, etc. have been plotted a freehand curve through these points will give the curve on the front surface of the block. Lines drawn across the top surface at 30 degrees from points o, 1, 2, etc. will give points o″, 1″, 2″, etc. on the back surface. Vertical lines dropped from these points and made the same length as those on the front edge will complete the curved portion of the block.

Application of solid geometry. A good knowledge of solid geometry is necessary to be able to set out any kind of joinery and carpentry work. A study of the various geometrical solids will enable the student to develop any shape he is likely to get in the building industry. The development of handrails, continuous staircase strings, louvre frames, work of double curvature, roofs to domes and turrets, are all based on geometrical solids.

Development of surfaces. To the left of Fig. 49 is the plan and elevation of a rectangular prism with its top surface a′–b′–c′–d′ inclined at an angle of 30 degrees. The plans are always kept below the X–Y line and the elevations above the X–Y line. The problems are to develop the shape of the top surface and also the four vertical surfaces. Draw the plan and elevation of the prism. Draw lines from a′ and b′ upwards and at right angles to a′–b′. Draw line a′–b′ on the development parallel to a′–b′ on the elevation, and make a′–d′ equal a–d in the plan. a′–b′–c′–d′ is the developed top surface.

To develop the vertical surfaces draw lines parallel to the X–Y lines from points a′ and b′ in the elevation. From any point a on the X–Y line mark off the four spaces a–b, b–c, c–d, and d–a equal to those round the plan. Draw vertical lines from the points obtained to intersect with the horizontal lines brought over from the elevation to give points a′, b′, c′, d′, and a′. Join these with straight lines to complete the development.

Fig. 50 is a similar problem but this time the solid is a hexagonal prism. The top surface has been developed in a downwards direction, the lines taken from the top inclined surface down to the developed surface are at right angles to the surface.

Fig. 51 is the plan and elevation of a square pyramid. To obtain the true length of corner o–a, place the compass point in o in the plan and with radius o–a describe an arc to give a″ on a horizontal line brought out from o. Project a vertical line upwards to give point a′ on the X–Y line. Join a′ to o in the elevation. a′–o is the true length of the corners of the pyramid.

To develop one side of the pyramid, with compass in b in the elevation and radius b–o, describe an arc to give o′ on the X–Y line. Drop a vertical line from o′ to intersect with a horizontal line brought out from o in the plan to give o″. Join o″ to c and o″ to b to obtain the shape of one side.

To develop the four sides of the pyramid (Fig. 52) with radius o–a′ describe an arc, and on the arc mark off four distances equal to those round the plan of the pyramid, to give points a′, b′, c′, d′, and a′. Join with straight lines as in the drawings and also join a to o, b to o, etc. with straight lines to complete the development.

Fig. 53 shows the plan and elevation of a hexagonal pyramid. The top portion of the solid has been removed leaving a surface inclined at 45 degrees. The inclined top surface can be developed by placing the compass point in d′ in the elevation and turning points c′, b′, and a′ round to the horizontal line brought out from d′. Drop vertical lines from these points on the horizontal line to intersect with horizontal lines brought out from the various points on the plan of the surface to give points on the development.

To develop the other surfaces of the pyramid it is necessary to obtain the lengths of the corners of the pyramid. As corners o–a and o–d in the plan are parallel to the X–Y line, their true lengths can be seen in the elevation. With the radius o–a in the elevation (the true length of all the corners of the pyramid) describe an arc (Fig. 54). Mark off, on the arc, six spaces equal to those round the plan of the pyramid. Join these with straight lines and to point o as seen in the drawing.

To obtain points a′, b′, c′, etc. project points b′ and c′ in the elevation over horizontally to the o–a line to obtain points b″ and c″. o–a′, o–b″, o–c″, and o–d′ are the true distances from the point of pyramid down to the corners of the top inclined surface. These distances should be marked on the appropriate lines in the development. o–e′ is equal to the distance o–c″, and o–f′ is equal to the distance o–b″.

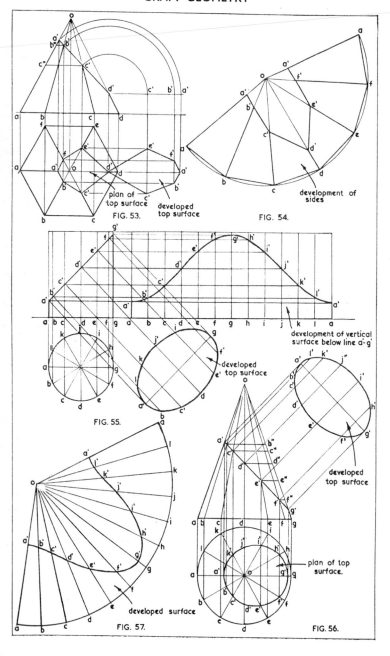

plan of top surface   developed top surface
FIG. 53.

development of sides
FIG. 54.

development of vertical surface below line a'- g'

developed top surface

FIG. 55.

developed surface
FIG. 57.

developed top surface

plan of top surface.

FIG. 56.

Fig. 55 is the plan of a cylinder with its top surface inclined at an angle of 45 degrees. Its top inclined surface and vertical surface can both be developed in a similar way to those in Fig. 50.

Fig. 56 is the plan and elevation of a cone. This solid, too, has been cut through at an angle of 45 degrees and the problems are the same as before, to develop the shape of the top inclined surface and also the surface of the cone from its base up to the top surface. To develop the shape of the top surface it is necessary, as with the pyramids, to obtain a plan of the surface. This is obtained by dropping vertical lines from the various points on the surface in the elevation down to the appropriate lines in the plan to obtain points a′, b′, c′ etc. A freehand curve through these points will give the plan.

To obtain the shape of the section, draw lines from all the points on the surface in the elevation at right angles to the surface to a position where the surface is to be developed. Draw the centre line a′–g′ parallel to the surface. Make lines b′–1′, c′–k′, etc. equal to those lines seen on the plan, marking the distances from the centre line each time.

To develop the surface of the cone up to the top inclined surface, the true lengths of the lines round the surface of the cone must first be ascertained. They all stretch from the apex o down to the base of the cone and therefore are all equal in length. With compasses open the distance o–a in the elevation (this is the length of all the lines round the surface of the cone) draw an arc (Fig. 57). Step off round the arc twelve spaces equal to those round the base of the cone, seen in the plan. Join these points with straight lines to o. This is the development of the whole of the cone.

To obtain points a′, b′, c′, etc. on the development the true distances down to these points must first be developed on the elevation. The distances o–a′ and o–g′ are true distances and can be transferred to the appropriate lines in the surface developments. Points b′, c′, d′, e′, and f′ must be projected horizontally across to one of the edges of the elevation to obtain their true distance down from the apex o. o–b″ is the true distance from o down to b′, o–c″ is the true distance down to c′, and so on. These distances should be transferred down to the correct lines in the surface development to be able to complete the drawing. The point 1′ is immediately behind b′ so the distance o–b″ will give the distance o–1′. The point k′ is immediately behind c′ so he distance o–c″ will give point k′ in the development, and so on. A freehand curve through all the points obtained will complete the drawing.

Also see chapter 33.

CHAPTER 6   SETTING OUT RODS

WHEN THE SCALE drawings or other details of a piece of joinery which has to be made have been received, the first job in the production of the work will be to produce what is called a workshop rod. This is simply a drawing, or a number of drawings, in full size which will give all the information necessary so that the job can be made.

In small firms which employ only a few joiners, it may be necessary for the person who is going to make the piece of joinery to prepare the rod himself, but in larger firms this job is usually left to one man or a team of men.   His or their job is to produce the rods for the joinery items which the firm has contracted to produce.   For economic reasons, it is obvious that the person whose duty it is to produce the rod must be skilled at his work.   He must be able to read the scale drawing and understand bills of quantities, have a good knowledge of joinery detail, and also produce a first-class drawing.

Rods are produced on narrow boards, sheets of plywood, or on rolls of setting-out paper, depending on the nature of the work. The three examples of setting-out rods in this chapter can be produced on any narrow material such as a 9-in.-wide board which has had its surfaces prepared and the edges shot straight.

Simple example.   Fig. 1 is the elevation of a small cupboard door, 3 ft. 0 in. high, 1 ft. 9 in. wide, and $1\frac{3}{8}$ in. thick.   It comprises two stiles; top, intermediate, and bottom rails (the bottom rail being wider than the others); and two plywood panels equal in size.   The rod must give the length, width, and thickness of the door, the sizes of all the framing members and position of the intermediate rail, the width and depth of the panel grooves, and the sizes of the panels.

To give all this information on the rod, two simple drawings are required; a horizontal section and a vertical section through the door. These can be seen on the rod illustrated in Fig. 2.   The vertical section, which gives the height of the door, gives also the positions and shapes of the three rails and the lengths of the two panels, and the horizontal section gives the width of the door, the shapes of the two stiles, and the width of the two panels.   Both drawings give the thickness of the door.

To produce a rod such as this, the person who is doing the setting out must try to imagine what he would see if he were to cut through

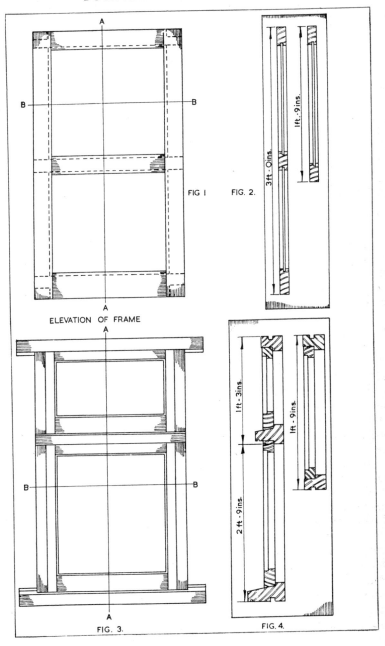

FIG I   FIG. 2.

ELEVATION OF FRAME

FIG. 3.   FIG. 4.

the door down the line marked A–A. If he has the imagination which all setters-out must have he will see the vertical section to the left of the rod. He must then imagine what he would see if he were to cut through the door in the other direction, on line B–B. He would, of course, see the horizontal section to the right of the rod.

Setting out a casement window. The second example involves a casement window-frame with transom. If the construction of the frame and casement below the transom is different from those above, two horizontal sections would be required on the rod. However, it should be quite simple to see that if the frame and casements were cut through in a horizontal direction on line B–B and also in a corresponding position above the transom, the pictures obtained would be the same in every detail. Consequently only one horizontal section is required.

The vertical section, to the left of the rod (Fig. 4), is what would be seen if the frame and casements were cut through on line A–A. It shows the shapes and positions of the head, transom, sill of the frame, and top and lower rails of both casements. The horizontal section through the work shows the position and shapes of the jambs of the frame and the stiles of the casements. It is usual, when placing the over-all dimensions on the rod, to clearly define the position of the more important parts, in this case the transom.

Door setting out. Fig. 5 is the elevation of a four-panelled door. As the upper portion of the door has a different construction from the lower portion it is necessary to show on the rod two horizontal sections. Figs. 6 and 7 show the two stages in the production of a rod for the door. In Fig. 6 only the outline of the three sections are shown. That on the left is the vertical section through line A–A; the upper right drawing is the horizontal section through line B–B above the middle rail; and that on the lower right of the rod is the section on line C–C below the middle rail. The object of the first stage, is of course, to make sure the over-all height and width of the door are correct, and that the finished sizes of the material and positions of the various members are in order.

When putting the smaller details on the rod, such as the shapes of the mouldings, it is a waste of time to draw them all, especially in such a case as the door in Fig. 5. The moulding is an ovolo moulding and would have to be drawn twenty-four times to complete the rod. This is obviously unnecessary, and probably one or two ovolos would suffice and the others finished as in the drawing. The over-all dimensions should be placed on the rod as well as the position of the middle rail.

ROD FOR ROOM DOOR

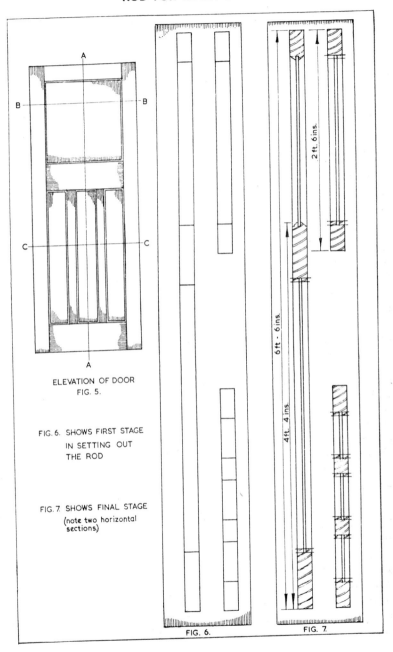

ELEVATION OF DOOR
FIG. 5.

FIG. 6. SHOWS FIRST STAGE
IN SETTING OUT
THE ROD

FIG. 7. SHOWS FINAL STAGE
(note two horizontal
sections)

FIG. 6.

FIG. 7.

CHAPTER 7   MARKING OUT THE MATERIAL

WHEN THE ROD has been completed the next step is to compile a list of the timber required to make the job.   An example of a timber list is given below and is for the cupboard door referred to in the chapter on setting out.

Cutting list—cupboard door (*see page* 85)

Item	Number required	Sawn sizes			Prepared sizes	
		Length	Width	Thickness	Width	Thickness
Stiles	2	3 ft. 2 in.	3 in.	$1\frac{1}{2}$ in.	$2\frac{3}{4}$ in.	$1\frac{3}{8}$ in.
Rails	2	1 ft. $9\frac{1}{2}$ in.	3 in.	$1\frac{1}{2}$ in.	$2\frac{3}{4}$ in.	$1\frac{3}{8}$ in.
Bct. rails	1	1 ft. $9\frac{1}{2}$ in.	4 in.	$1\frac{1}{2}$ in.	$3\frac{3}{4}$ in.	$1\frac{3}{8}$ in.
Panels	2	1 ft. 2 in.	1 ft. 4 in.	$\frac{1}{4}$ in. ply		

Once the timber has been prepared to size, the rod and the timber are sent to the person who is to mark out the work.   This includes the placing of the face marks as well as marking the position of the mortises, tenons, rebates, etc.   This work, in small joiners' shops, is also done by the joiner himself.   The tools commonly used for marking out are the marking and mortice gauges, try square, sliding bevel, marking knife, and pencil.   The three jobs referred to in the last chapter are also taken as examples to illustrate the methods used in the marking-out of timber.

Cupboard door example.   For the first example let us take the small cupboard door, Fig. 1, Chapter 7.   Figs. 1 and 2 show how two of the rails for the job are marked out.   The rails are a little longer than 1 ft. 9 in. so that the ends of the tenons will protrude through the mortises, allowing the ends to be trimmed and cleaned off after the frame has been assembled.

The best face and edge should then be selected from each piece of timber and on these surfaces should be placed the face-side and face-edge marks (see Fig. 2).

After checking that the horizontal section is 1 ft. 9 in. wide, the three rails in turn can be placed on the horizontal section of the rod and the positions of the shoulders made on the face edge.   With the use of the square and marking knife these marks should be squared round the four surfaces of the timber.   The mortice gauge is set to suit the thickness of the mortice and tenon being used, and the

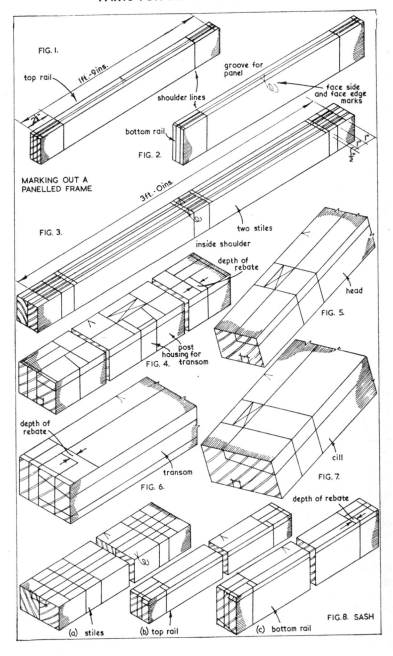

FIG. I.

top rail

1ft.-9ins.

groove for panel

shoulder lines

face side and face edge marks

2"

bottom rail

FIG. 2.

1"
1"
½

MARKING OUT A PANELLED FRAME

FIG. 3.

3ft.-0ins.

two stiles

inside shoulder

depth of rebate

head

FIG. 5.

post housing for transom

FIG. 4.

depth of rebate

cill

FIG. 7.

depth of rebate

transom

FIG. 6.

(a) stiles

(b) top rail

(c) bottom rail

FIG. 8. SASH

79

HOW NOT TO TREAT JOINERY
These frames are exposed to the weather and badly stacked leading to various sorts of damage.

CASEMENT FRAME HELD IN POSITION BY A SCAFFOLD BOARD
Metal clamps are screwed on the sides of the frame as the brickwork proceeds.

tenons marked round the ends of each piece with the fence of the mortice gauge against the face side of the timber.

Another mortice gauge is set to the thickness of the panel being used, and the panel groove marked on the face-edge surface of each piece, with the fence of the gauge again against the face side of the timber. One of the pieces to be used as stiles should be placed on the vertical section of the rod after checking that the section length is 3 ft. 0 in. long, and the positions of the mortices and haunches marked on the face-edge surface. The mortice positions can be squared round to the back edge with square and pencil.

It must be remembered that the two stiles must be marked out as a pair (see Fig. 3), and note should be made that the face side of one piece is facing away, and that of the second piece is facing forwards. The two pieces should be placed together as seen in the drawing, and the marks made on the piece when it lay on the vertical section of the rod can be transferred to the second stile. This will ensure both stiles being marked out exactly the same.

The two mortice gauges, set up previously for marking the tenons and grooves on the rails can be used for marking the mortices and grooves on the stiles, remembering to keep the fences of both on the face side of each stile.

Marking out casements. Figs. 4–7 show how the members forming the frame for the casements in Fig. 3, Chapter 7 are set out. Fig. 4 is a shortened view of one of the posts of the frame, and as there are two of them they should be set out as a pair as previously described. Note here that allowances have to be made for the shoulders on the rebate side of the frame. If the rebate is half an inch in depth the shoulders on this side of the posts must be half an inch longer than those on the opposite side. When marking the shoulder on the rebate side at the sill ends of the posts remember to allow for the weathering on the sill.

The head (Fig. 5) and sill (Fig. 7) are set out similarly, but, the sill being wider than the head, it follows that the rebate will be wider and, of course, the rebate has to be bevelled to allow for weathering. The transom (Fig. 6) has two rebates; the upper one is weathered similarly to the sill, and the lower one is square. Allowances must be made for the rebates in the posts and the housing in the front edges of the posts.

Fig. 8 shows how the stiles and rails for the casements are marked out. Again, the stiles must be set out as a pair, and when marking out the rails the shoulders on the rebate sides must be lengthened by the depth of the rebates.

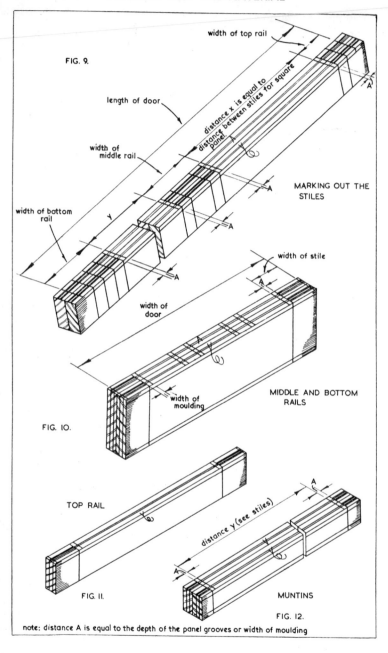

FIG. 9.

width of top rail

length of door

distance x is equal to distance between stiles for square panel

width of middle rail

MARKING OUT THE STILES

width of bottom rail

width of stile

width of door

width of moulding

MIDDLE AND BOTTOM RAILS

FIG. IO.

TOP RAIL

distance y (see stiles)

FIG. 11.

MUNTINS

FIG. 12.

note: distance A is equal to the depth of the panel grooves or width of moulding

Four-panelled door. Figs. 9–12 show how the four-panelled door in the setting-out chapter is marked out. One of the stiles should be laid on the vertical section of the setting-out rod after checking the length of that section, and the position of the three rails marked on the face-edge side. The depth of the panel grooves is marked back from the lower edge of the top rail, both edges of the middle rail, and top edge of the bottom rail (distance A). The remaining distances in each case are divided up in the correct proportions for the mortises and haunchings. These marks are squared round with a pencil to the other edge of the stile.

Two mortice gauges, one set to the thickness of the tenons, and the other to the thickness of the panels, are used with the fence on the face side. The other stile should be placed alongside, remembering that both are to be marked out as a pair, and the position of the rails on the first stile squared over to the face edge of the second one.

The top rail (Fig. 11) and the middle and lower rails should be placed on the lower horizontal section of the rod, and the position of the inside edges of the stiles placed on the face edges. Remember that allowances must again be made, this time for the mouldings to be worked on all the inside edges of the frame. When marking the position of the shoulders to the middle and bottom rails, the positions of the muntins, too, should also be placed on the appropriate edges of these two members.

To avoid any trouble here it is as well to set the middle and bottom rails out as a pair so that the positions of the mortises for the muntins will coincide on each piece. One of the muntins can then be placed in the correct position of the vertical section of the rod, and the edges of the middle and bottom rails marked on the face edge. Allowances, as before, must be made for the mouldings.

One important thing before closing this chapter is that face-side and face-edge marks must always be placed on joinery items, and shoulder lines must be marked with a knife or other sharp tool.

SIMPLE FORMWORK FOR CONCRETE WORK
Photograph by courtesy of the Cement and Concrete Association.

FORMWORK USED IN MAKING A SWIMMING POOL
Photograph by courtesy of the Cement and Concrete Association.

CHAPTER 8   FORMWORK FOR CONCRETE

CONCRETE IS USED to a great extent in building, and carpenters who have to carry out the work of making the boxes or forms into which the concrete is placed have to be skilled in this work if the efforts of their labour are to produce the desired results.   Concrete products can either be cast *in situ*, which means that they are actually made in the position for which they are intended; or they can be pre-cast, in which case they are made in any convenient position, and moved to the site when set.

It is a much more practical job to pre-cast such items as kerbstones or paving slabs for a pathway and deliver them on the site when they are required, but when it comes to a job such as a concrete bed for a piece of machinery this would be cast *in situ*.   There are, of course, instances where the work could be done either way, such as a concrete lintel over a doorway or concrete window-sills, and conditions at the time would decide on which method would be used. If, say, a large number of lintels for garage doors were required for a building site, it would be more economical to make a small number of boxes for them to be cast at the building firm's works (assuming they were to manufacture the lintels themselves) than it would be to cast one lintel at a time *in situ* as each garage is built.   In a case such as this, half a dozen boxes would be made so that six lintels could be cast at one time, and the boxes cleaned and used again for the next batch, and so on.

Considerations.   Concrete is a heavy material, and it is important to remember when making formwork and supporting it that the formwork and the supports must have sufficient strength to withstand the weight.   Provision must be made also, where necessary, to ensure that the timber formwork and the supports can be removed fairly easily.   On big jobs the safety of the workmen, too, must be kept in mind at all times.

The design of the formwork should be kept as simple as possible, but the finished product must be exactly as the drawings or the specification demand.

Simple paving slab.   Fig. 1 is the formwork required to produce a paving slab 2 ft. 6 in. long, 2 ft. 0 in. wide and 2 in. thick. It comprises four pieces of 2 in. by 2 in. softwood, the shorter pieces

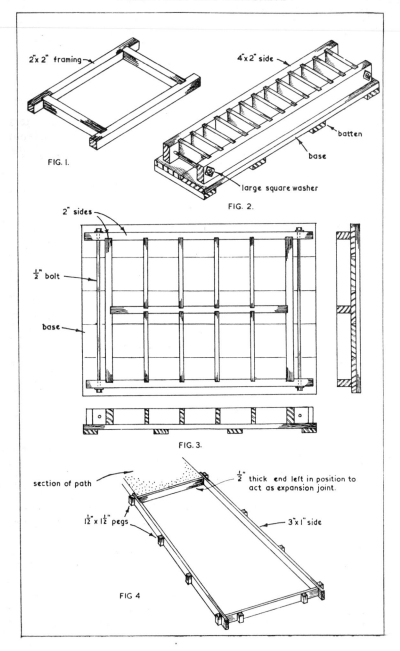

2"x 2" framing

4"x 2" side

batten

base

FIG. I.

large square washer

FIG. 2.

2" sides

½" bolt

base

FIG. 3.

section of path

½" thick end left in position to act as expansion joint.

1½" x 1½" pegs

3"x 1" side

FIG 4

being housed into the other two sides to a depth of about $\frac{1}{2}$ in. To prevent the longer sides from being pushed outwards by the concrete some heavy blocks could be placed against them. If a more elaborate form of box is required the sides could be secured together by means of long bolts, and a base board could also be provided each as detailed in Fig. 2.

This drawing illustrates how twelve concrete blocks 9 in. by $4\frac{1}{2}$ in. by 3 in. thick can be pre-cast. The box consists of two long sides made from 4 in. by 2 in. softwood, and housings are made in each of these sides to receive the $4\frac{1}{2}$ in. by 1 in. divisions. Two bolts are used to keep the box together until it is necessary to remove the blocks. A base board, made from 1 in. material and well battened together, should be placed on a site which has been carefully levelled before filling commences.

Large blocks. A box for pre-casting larger blocks is shown in Fig. 3. In this case the box provides for ten blocks to be made at a time, and consists of four sides and a centre division which divides the box up into two halves. Each half has four smaller divisions. The box is held together with two $\frac{1}{2}$ in. bolts, and the whole rests on a base board made in a similar manner to that in Fig. 2. Two horizontal sections through the box are also shown.

If path or larger slabs have to be made these can be cast *in situ* in a similar manner to that shown in Fig. 4. The ground on which the slab is to be cast should first have the vegetable soil removed from it and hardcore placed in the excavated portion and rammed down hard. A four-sided box with its inside dimensions equal to those of the required slab should then be prepared with 1-in.-thick boards. When positioned correctly it should have their top edges level. The corners of the box can be lightly nailed together but the rigidity of the box will rely on square pegs driven into the ground at about 2 ft. centres. These will prevent the concrete, when poured into the box, from pushing the sides of the box outwards. The length of each side of the concrete slab should not be more than 10 ft. If it is required to be larger it must be made up in a number of sections so that the concrete will be able to expand or shrink. Then strips of timber can be left in between the slabs to act as jointing strips.

Posts. Concrete posts suitable for chain link fences can be pre-cast in a box similar to that shown in Figs. 5, 6, and 7. Fig. 6 is the plan of the box and shows that four posts can be cast in one operation. Fig. 5 is the elevation of the box. The posts to be produced are 4 in. by 4 in. and are to be weathered at the top ends and have four holes through each through which wires can be passed.

The two long sides of the box and the three divisions are made from 4 in. by $1\frac{1}{2}$ in. softwood, and the two ends from 4 in. by 2 in. material. The weatherings are formed by placing triangular pieces of wood in one of the top corners of each section. The four holes are formed in each of the posts by having four $1\frac{1}{2}$ in. by $1\frac{1}{2}$ in. battens fixed across the box with four $\frac{1}{2}$ in. holes bored in each as shown in the plan. Through each of the holes is placed a $\frac{3}{8}$ in. piece of mild steel, and the lower ends of these enter a shallow $\frac{1}{2}$ in. hole bored in the base board on which the box stands.

To avoid the imprints of the four battens being made on the surfaces of the posts, the battens are raised up slightly above the top surface of the box by fixing battens along the two long sides and on the centre division, and the four battens which hold the mild steel rods screwed to these.

Fig. 7 shows, to a larger scale, a section through the box and shows how the battens which hold the mild steel rods are raised above the top surface of the box. The battens also help stop the sides of the box from being pushed outwards by the concrete.

Steps. Figs. 8, 9, and 10 show how a couple of concrete steps can be cast *in situ*. They would be suitable for leading up to an entrance door to a shed or house.

Two shutters are first made, the shapes being equal to the outline of the steps. These are fixed in position by first plugging two battens on the face of the wall and then nailing the shutters to the battens. The riser boards should be nailed to the front surfaces of the shutters. For easy removal of the nails the heads should be left protruding about $\frac{1}{4}$ in. So that the lower edge of the top riser board will not be imprinted on the lower concrete step its lower edge should be bevelled as shown in Fig. 11.

Lintel. Figs. 12 and 13 illustrate how a 9 in. by 9 in. concrete lintel over a garage doorway can be cast *in situ*. A box with two sides and a bottom is first constructed, making the inside dimensions 9 in. by 9 in. The bottom of the box should fit between the sides for maximum strength and should be cut in length to fit between the jambs of the brickwork. Its thickness should be $1\frac{1}{2}$ in. to 2 in. The sides of the box should be $1\frac{1}{4}$ in. to $1\frac{1}{2}$ in. thick, and their ends should project approximately 12 in. beyond the ends of the bottom (see Fig. 16). The reason for this is seen in Fig. 13—the lintel has to have a 9 in. bearing on the brickwork. To prevent the sides from being pushed outwards by the weight of the concrete, distance pieces are placed over the top of the box at about 3 ft. centres (see Fig. 14).

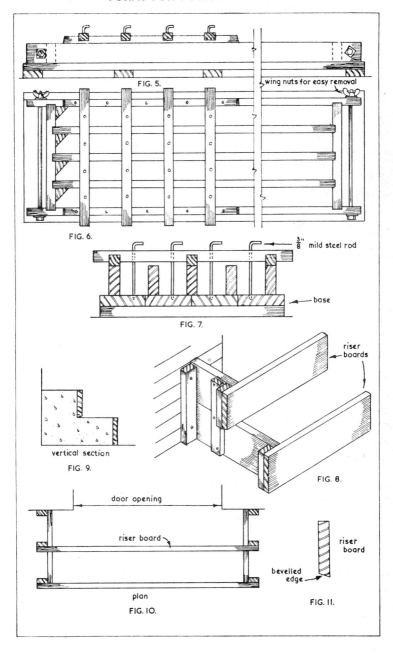

FIG. 5.

wing nuts for easy removal

FIG. 6.

$\frac{3}{8}$" mild steel rod

base

FIG. 7.

vertical section

FIG. 9.

riser boards

FIG. 8.

door opening

riser board

plan

FIG. 10.

riser board

bevelled edge

FIG. 11.

FORMWORK FOR CONCRETE

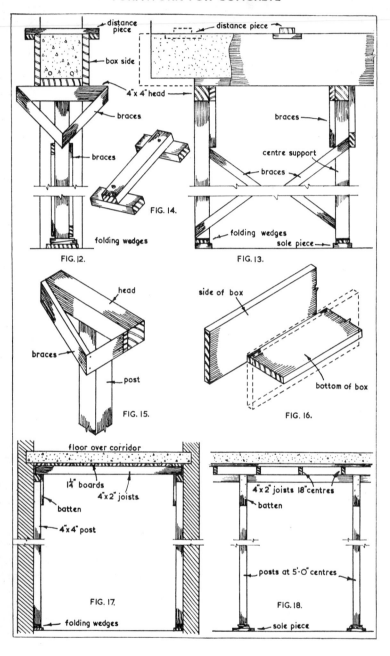

distance piece

distance piece

box side

4" x 4" head

braces

braces

braces

braces

centre support

braces

folding wedges

FIG. 14.

folding wedges

sole piece

folding wedges

FIG. 12.

FIG. 13.

head

side of box

braces

post

bottom of box

FIG. 15.

FIG. 16.

floor over corridor

1¼" boards

4"x 2" joists

batten

4"x 4" post

4"x 2" joists 18" centres

batten

posts at 5'-0" centres

FIG. 17.

folding wedges

FIG. 18.

sole piece

The box has to be supported about 7 or 8 ft. above the ground and this is done by constructing supports similar to that shown in Fig. 15. A post, out of 3 in. by 3 in. or 4 in. by 4 in., and a head of similar dimension and about 1 ft. 6 in. long are placed together as shown in the sketch, and are kept at right angles to each other by means of braces.   Under each post should be a small piece of timber called a sole piece, and, to enable final adjustments to be made and also to allow easy striking of the formwork, a pair of folding wedges is placed between them.   The supports can first be placed in position and braced as shown in the elevation.   The box is lifted up and placed on the heads of the supports.   It can be fixed in position by nailing through the bottom of the box into the heads of the supports.

The folding wedges are used to raise the box up to its correct position, and when the time arrives for the formwork to be struck the folding wedges can once again be used to release the woodwork which would otherwise be very firmly wedged under the weight of the concrete.

Corridor floor slabs.   The next illustrations show how a simple floor slab over a corridor can be formed.   Boards $1\frac{1}{4}$ in. thick running in the direction of the corridor are supported by 4 in. by 2 in. joists at 18 in. centres.   These in turn are supported by 4 in. by 4 in. heads running in the same direction as the boards and have posts spaced at 5 ft. centres supporting them (see Figs. 17 and 18).   To prevent the posts from becoming displaced at their tops battens are nailed across their inside surface and to the heads.   Sole pieces and folding wedges are again used to make final adjustments in the height of the formwork and also to assist in its easy removal.

Concrete floor on steel beams.   Fig. 19 illustrates how the formwork is made for a concrete floor which is supported by steel beams.   The concrete is supported by $1\frac{1}{4}$-in.-thick softwood boards, and these are supported on 6 in. by 2 in. joists spaced at 18 in. centres.   The sizes of the joists must be increased as the span increases.   The wall end of the joists are supported in a similar manner as in Figs. 17 and 18.   The beam box has to be made to the requirements, and this is supported by posts and heads made in a similar way to those which support the lintel box in Figs. 12 and 13. The posts and heads should be spaced along the length of the beam at about 5 ft. centres.   The ends of joists at the beam end are carried by a bearer which rests on the heads and it can also be fixed by screwing to the beam-box sides.   Folding wedges under all the posts again will bring the top surfaces of the boards up to the correct

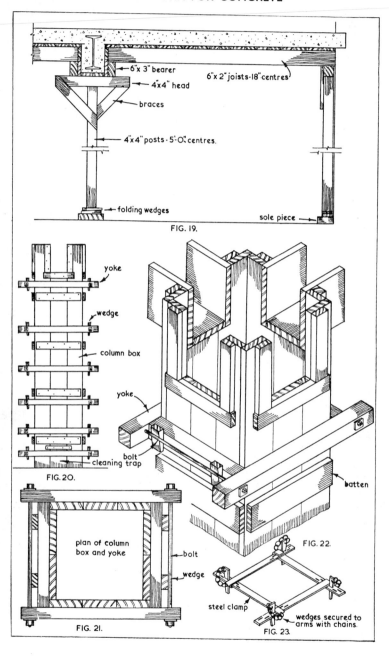

FIG. 19.

6"x 3" bearer
4"x 4" head
braces
4"x 4" posts - 5'-0" centres.
6"x 2" joists - 18" centres
folding wedges
sole piece

FIG. 20.

yoke
wedge
column box
yoke
bolt
cleaning trap

FIG. 21.

plan of column box and yoke
bolt
wedge

FIG. 22.

batten

FIG. 23.

steel clamp
wedges secured to arms with chains.

level, and will also assist in the striking of the work when the time for dismantling arrives.

Columns. Concrete floors of extra large span usually have to be supported by columns spaced across the spans in each direction in equal intervals. The columns support beams which, in their turn, support the various sections of the floor. The formwork required for a square concrete column is shown in Fig. 20. It comprises four shutters well battened together, and when placed together to form a four-sided box the internal dimensions should equal the size of the concrete column.

Fig. 21 is a horizontal section through the box. The box is held securely together by a series of what are called yokes. Each yoke consists of four fairly strong pieces of timber, bolted round the column box at intervals throughout its entire length as shown in Fig. 21. The bolts should have large square washers behind the heads and the nuts to prevent their being crushed into the woodwork when the concrete is poured into the box. Wedges are placed between the bolts and shorter sides of the yokes to keep them tight up against the sides of the box.

It is necessary to keep the yokes fairly close together at the bottom of the box because of the great weight of the concrete when it is in a semi-liquid state. A good rule is to keep the bottom yoke about 6 in. from the floor level and increase the distance between each pair by 3 to 4 in.

Often steel yokes or cramps are used instead of timber ones, and these are much more serviceable as they can be adjusted more easily and are secured in position with metal wedges (see Fig. 23).

It is necessary to leave a small trap at the bottom of the box so that the space inside the box can be cleared of short ends of timber and any other items which may be accidently dropped when the erection of the formwork is taking place. The top of the box will also have to be prepared to take the ends of the beam boxes which intersect at the columns. Details of this arrangement are shown in Fig. 20 and also in the isometric view (Fig. 22). To strengthen these points and to give the ends of the beam boxes a better bearing, battens are screwed round the openings for the beams. The beam boxes are supported between the columns in a similar way to that shown in Fig. 19. The supports should be spaced at about 5 ft. centres.

Also see chapter 22.

CHAPTER 9   CENTRES FOR ARCH WORK

TEMPORARY FRAMES HAVE to be made from time to time to assist the bricklayer and stone mason to construct arches over openings in buildings. It is the work of the carpenter to make these frames but before he makes them he must know how to set them out.

Arches. Fig. 1 is the elevation of a segmental arch. a–b is the clear span, and c–e the rise. To draw the curve of the arch draw a–b and c–e. Bisect a–c. The bisecting line will give point d on the extended centre line. With compass in d and radius d–a draw the curve a–c–b.

Fig. 2 is the elevation of a semi-circular arch, and should need no explanation.

Fig. 3 is an equilateral arch with a clear span a–b. With compass point in a and b in turn and with radius a–b, describe the two arcs a–c and b–c. Point c is on the centre line.

Fig. 4 is the elevation of an arch of approximately elliptical form, and is drawn with the aid of compasses. This is used for gauged brick arch work. Draw the span a–b and the rise h–c. Draw the diagonal a–c, and with centre h and radius h–a describe the arc to give point d on the extended centre line. With centre c and radius c–d describe the arc to give point e on the diagonal a–c. Bisect a–e, the bisecting line will give point 2 on the line a–b and point 1 on the centre line. Make h–3 equal h–2, and draw the line from 1 through point 3 to pass through point g. Using the centres 1, 2, and 3 the curves a–f, f–g and g–b can be drawn. These form the outline of the arch.

In gauged brickwork the bricks are shaped similarly to those shown round the outside of the curve already drawn. It is the work of the carpenter to produce the templets for the bricklayer to use for shaping the bricks. If the elliptical arch outline in Fig. 4 has been drawn with the aid of compasses only two templets will be required because the bricks which come immediately over the curves a–f and b–g will all be the same shape, and the bricks above the curve f–g will also be one shape. The two templets are shown shaded on the drawing.

Tudor arch. Fig. 5 is the elevation of a Tudor arch. Let a–b be the span and c–d the rise. Construct the rectangle a–d–c–e

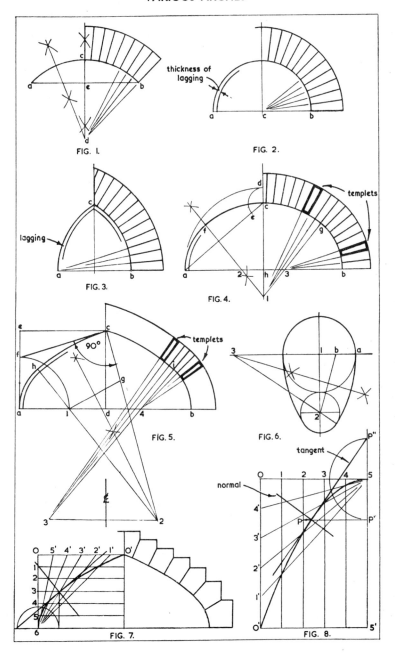

FIG. 1.

FIG. 2.

thickness of lagging

FIG. 3.

lagging

FIG. 4.

templets

FIG. 5.

templets

90°

FIG. 6.

FIG. 7.

FIG. 8.

tangent

normal

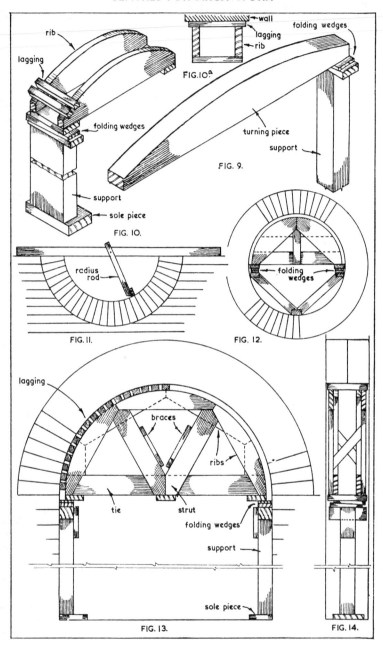

rib

lagging

wall

lagging

rib

FIG.10.ᵃ

folding wedges

folding wedges

support

turning piece

support

FIG. 9.

support

sole piece

FIG. 10.

radius rod

folding wedges

FIG. 11.

FIG. 12.

lagging

braces

ribs

tie

strut

folding wedges

support

sole piece

FIG. 13.

FIG. 14.

as shown, and divide a–e into three equal parts. a–f is two-thirds a–e. Join f to c and angle f–c–2 a right angle. Make a–1 equal a–f and c–g also equal a–f. Bisect 1–g to give point 2 below the line a–b. Draw a line from 2 through point 1.

With centre 1 and radius a–1 describe the arc a–h. With centre 2 and radius 2–h describe the arc h–c to complete one half of the outline. The second half is similar to the half already drawn. Two templets are also required for the bricklayer if this arch is constructed with shaped bricks.

The arch outline in Fig. 7 is a pointed parabolic arch. The curve should be set out in a similar way to the parabola shown in Fig. 8. Point o' at the base of the parabola is at the top of the rise of the pointed arch in Fig. 7. To construct the parabola, Fig. 8, first construct the rectangle, o–o'–5–5' to any required size and divide o–5 and o–o' into the same number of equal parts. Draw vertical lines downwards from the points on o–5, and draw lines from the points on o–o' to intersect with point 5. 1' and 1 intersect to give a point on the curve of the parabola, 2' and 2, 3' and 3, etc. give the other points required through which to draw a freehand curve.

To construct a tangent to pass through point P on the curve, draw a horizontal line across to 5–5' to give point P'. With centre 5 and radius 5–P' describe an arc to give point P" on the extended 5'–5 line. Draw a line from P" to pass through P on the curve. This is the tangent required and a normal to pass through the same point is drawn at 90 degrees to the tangent.

To construct the egg-shaped figure in Fig. 6 draw the two circles with centres 1 and 2 to any required size. Make a–b equal the radius of the small circle and then bisect b–2. The bisecting line will give point 3 on the extended diameter line of the large circle. Centre 4 is found in the same way as 3 was found.

Temporary arch supports. Figs. 9 and 10 show temporary supports for the brickwork of segmental arches. When the brickwork is only 4½ in. in thickness the turning piece in Fig. 9 is all that is required. It is cut to the curve of the arch and can be obtained from a 3 in. piece of timber. It is supported by a piece of timber at each end, and folding wedges are placed between the turning pieces and each support so that small adjustments can be made. The wedges also assist in the striking of the centres when they have to be removed.

The segmental arch centre in Fig. 10 is suitable for a wall which is more than 4½ in. thick. Before cutting the shaped ribs it must first be decided what thickness lagging is going to be used for the job. Lagging consists of small battens about ¾ in. by ¾ in. or ¾ in. by 1 in.

which are nailed round the top edges of the ribs to form a platform for the bricks to be placed upon.

When setting out the arch two curves are drawn, the first being the outline of the arch, and the second a line parallel to the arch outline but the thickness of the lagging away. It is the inside curve which give the shapes of the ribs (see Figs. 1–5). The ends of the lagging should not extend to the faces of the wall (see Fig. 10a). For a 9 in. wall the laggings should be approximately 8 in. long, and when the pieces are fixed to the ribs they should extend beyond the faces of the ribs to a distance of about $\frac{1}{2}$ in. at each end (see also Fig. 10a). The lower edges of the ribs are secured by nailing short ends of battens across their lower surfaces. These also assist in the proper adjustment of the folding wedges.

Circular openings. Bullseye or circular openings are formed in two stages. Fig. 11 shows the first stage, and for this the carpenter has to supply a radius rod and a batten to fix it so that it can be turned in a full semi-circle. The radius rod assists the bricklayer in laying the arch bricks in the lower half of the opening. The second stage, Fig. 12, is carried out by providing the bricklayer with a small semi-circular centre so that the upper half of the bullseye opening can be constructed. This centre has to be propped up in the lower half of the opening with short pieces of timber, using two pairs of folding wedges inserted to assist in the final adjustments and easy removal as seen in the drawing.

Large semi-circular opening. Figs. 13 and 14 are the elevation and vertical section through a larger semi-circular centre. The ribs are built up in two thicknesses, the joints being staggered and the pieces nailed together. The nails should be long enough to allow their ends to be clenched. A tie is used for securing the ends of the built-up ribs, the front thicknesses overlapping the front surface of the tie at each end and the two securely clench-nailed together.

Two struts are also employed to strengthen the ribs, these being cut to fit in the angle formed by two of the front ribs and nailed to the rear thickness ribs. The lower ends of the struts are nailed to the front surface of the tie. The centre is supported in a similar way to that in Fig. 10.

Other arch centres. Fig. 15 is the elevation of an equilateral arch centre and is constructed and supported similarly to the semi-circular centre. Fig. 16 is the side elevation of the centre and shows the lagging secured to the outside surfaces of the ribs.

Fig. 17 is the elevation of an approximate elliptical arch centre, and where a fairly wide arch is involved it is usual to include three

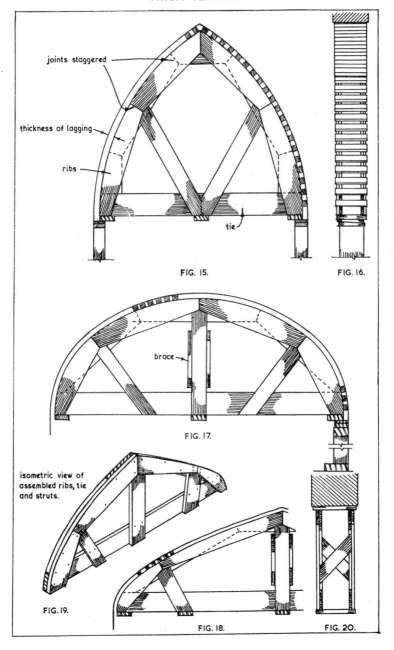

joints staggered

thickness of lagging

ribs

tie

FIG. 15.

FIG. 16.

brace

FIG. 17.

isometric view of
assembled ribs, tie
and struts.

FIG. 19.

FIG. 18.

FIG. 20.

struts, one on each side of the centre line, and one positioned centrally. So that the completed centre cannot become out of vertical alignment, cross or diagonal bracing is advised. This can be seen in the elevations of Figs. 13, 17, and 18, and the vertical sections of Figs. 14 and 20.

Fig. 18 is part elevation of a centre suitable for a Tudor arch, and Fig. 19 shows clearly how the ribs for this centre and also for the previous ones are built up. As already stated, they are built up of two thicknesses with the joints staggered and the pieces well nailed together. The tie secures the feet of the ribs and so stops any tendency for them to spread. The struts are nailed to the tie and the back layer of ribs. Fig. 20 is a vertical cross-section through the Tudor arch and shows the diagonal bracing nailed to the centre strut. *Also see chapter 20.*

CHAPTER 10   GROUND FLOORS

ONE OF THE first jobs a carpenter has to do on a building site is to construct the ground floor of the first house being built. The bricklayers will bring the brickwork up to the required level, and the carpenter will then take over and complete his part of the work before the bricklayers can proceed with their job.

Fig. 1 is the plan of the timber in a typical ground floor. The joists to which the floor-boards are nailed are spaced evenly across the length of the room, usually about 16 to 18 in. centre to centre. To enable fairly narrow joists to be used, honeycombed sleeper walls are built up from the surface concrete so that the amount of clear span for each joist will not exceed 5 ft. Figs. 2 and 3 illustrate the sleeper walls. The honeycombing or spaces between the bricks in the walls allow a free passage of fresh air (which is supplied from outside by the air bricks) to pass over all the surfaces of the timbers and so prevent them from absorbing moisture beyond 20%, which is the level of moisture required by the fungus dry rot before it can attack timber.

Damp-proof courses. Before the carpenter proceeds with his work he should make certain that damp-proof courses of bitumen felt or similar material have been placed on the sleeper walls, the internal walls, and the fender wall which is around the fireplace opening. These damp-proof courses are necessary, of course, to prevent moisture rising up from the ground, through the foundations and brickwork, to the timbers in the floor. Timber plates should be bedded in cement mortar on the sleeper walls and the front portion of the fender wall to give a fixing to the joists.

Often a timber plate is omitted from the partition walls of the house, this space being taken up with another course of brickwork, the ends of the joists resting on these bricks (Fig. 2). The joists, usually 4 in. by 2 in. in size, are placed over the wall plates on the sleeper walls and brickwork on the internal walls, and spaced evenly across the length of the room.

For 1 in. floor-boards the joists should not be more than 16 in. apart or 18 in. centre to centre. The end joists should be about 2 in. from the walls. When positioned correctly, the joists can be securely nailed to the plates on which they rest.

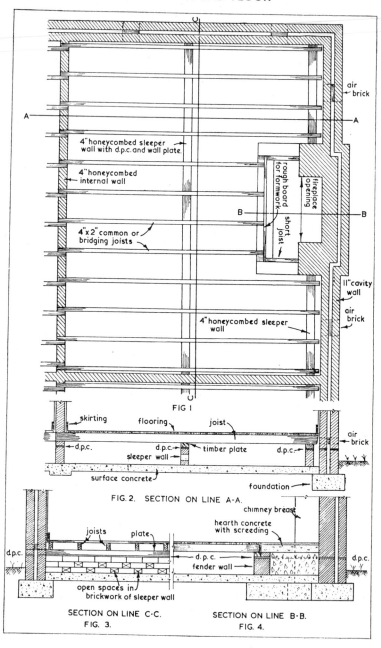

air brick

A ———— A

4" honeycombed sleeper
wall with d.p.c. and wall plate.

4" honeycombed
internal wall

4" x 2" common or
bridging joists

rough board for formwork

fireplace opening

B ———— B

short joist

11" cavity wall

air brick

4" honeycombed sleeper
wall

FIG I

skirting   flooring   joist

d.p.c.   d.p.c.   timber plate   d.p.c.

air brick

sleeper wall

surface concrete

foundation

FIG. 2. SECTION ON LINE A-A.

chimney breast

hearth concrete
with screeding

joists   plate

d.p.c.   d.p.c.   d.p.c.

fender wall

open spaces in
brickwork of sleeper wall

SECTION ON LINE C-C.
FIG. 3.

SECTION ON LINE B-B.
FIG. 4.

FIREPLACE DETAIL

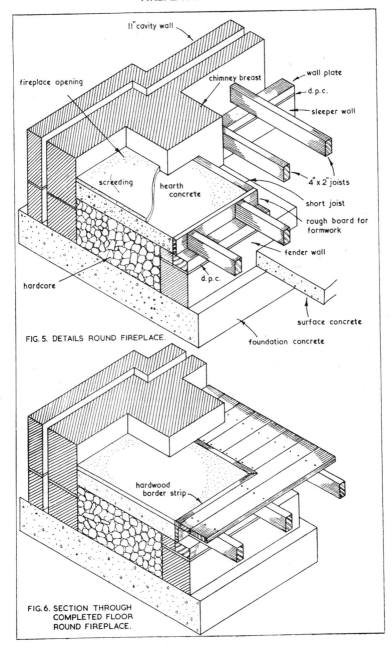

11" cavity wall

fireplace opening

chimney breast

wall plate

d.p.c.

sleeper wall

screeding

hearth concrete

4" x 2" joists

short joist

rough board for formwork

fender wall

hardcore

d.p.c.

surface concrete

foundation concrete

FIG. 5. DETAILS ROUND FIREPLACE.

hardwood border strip

FIG. 6. SECTION THROUGH COMPLETED FLOOR ROUND FIREPLACE.

Hearth. The space between the fender wall and chimney breasts is filled with hardcore up to the top of the fender wall, and a 4-in.-thick concrete hearth is cast, *in situ*, over the hardcore. To enable the concrete to be placed in this position a short length of rough boarding, the same depth as the joists, should be nailed to their ends as shown in Fig. 1, and two short 4 in. by 2 in. joists fixed across the width of the hearth in the position shown in the drawing, to act as formwork for the concrete. The concrete is poured into the hearth space and brought up to the same level as the floor-joists (see Fig. 4).

Floor-boards. When the floor-boards have been laid (this should be delayed until the plastering has been completed if possible), the 1-in.-thick sand and cement screeding is placed over the concrete hearth to give it a smooth finish. It is necessary to delay the fixing of the floor-boards as long as possible, at least until after the roof has been completed, because with constant traffic over them during building operations, their surfaces would become very dirty and possibly damaged. If possible, the boards should be cut and laid face side downwards so that work can proceed, in the normal way. Later they are turned over and fixed.

It is also necessary to see that the floor-boards are adequately protected when they are delivered on to the site, so that when they are fixed undue shrinkage will not occur. They should be unloaded from the lorry and placed in the shed provided for storing joinery items or, if this is not available, they should be carefully stacked in one of the houses which has had its roof completed.

Fig. 5 is an isometric picture of part of a ground floor around a fireplace. The 11 in. external cavity wall is shown; also portions of the concrete foundations, surface concrete, and fender wall. Note that damp-proof courses are placed in all the brickwork below floor level. The 4 in. by 2 in. floor-joists are seen resting on wall plates, and the method of forming the hearth with a piece of rough boarding and one short joist is also clearly seen. The cement screeding is laid over the concrete after the floor-boards have been fixed.

Fig. 6 shows the same portion of the floor when it has been completed. A hardwood strip, equal in thickness to the floor-boards and mitred at the corners, is often placed round the three edges of the hearth to give a good finish.

CHAPTER 11   UPPER FLOORS

IN CONTRAST TO those of a ground floor, the joists which form a single upper floor to a domestic house normally have no intermediate supports. They have to span the whole of the space to be occupied by the floor, resting on the brick or partition block walls at their ends. As the span for one of these floors can be anything up to 16 ft. or more, much greater depth of joist is required. How can one decide on an economical size of joist which will be able to support the loads to which it may be subjected? There are two ways to do this; the rule-of-thumb method which is not always to be recommended; and that involving careful calculation, which is the more desirable method of obtaining the dimensions of the required joists.

Rule-of-thumb calculation.   In this it is necessary to take the clear span of the floor in feet, divide this figure by two, and then add two, the result being the depth of the joists in inches.   For example, let us assume that the clear span of a floor space is 10 ft.:

$$\frac{10}{2} + \frac{2}{1} = 7 \text{ (in.)} = \text{depth of joist}$$

As the common thickness for joists is 2 in., a quantity of 2 in. by 7 in. joists would be suitable for a floor having a 10 ft. span.

Joist calculations.   The second method is more involved. Several factors are first required before the formula which will give the sizes of a joist which is to carry a uniformly distributed load can be calculated.

The formula is:

$$\frac{Wl}{8} \text{ (the bending moment)} = \frac{fbd^2}{6} \text{ (the moments of resistance)}$$

W = uniform load to be carried by the joist, total,
l = span of joist in inches,
f = fibre stress of timber per square inch,
b = breadth of joist,
d = depth of joist.

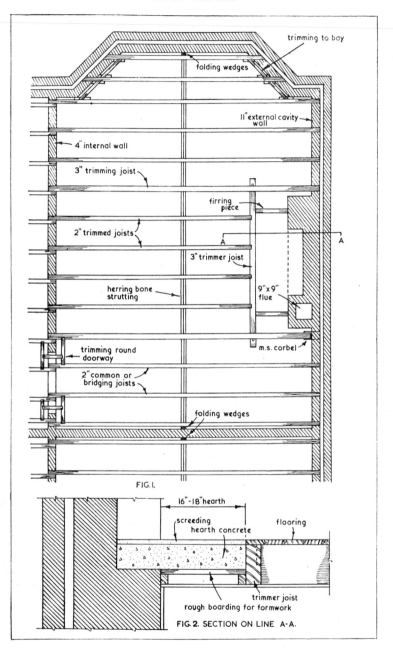

trimming to bay

folding wedges

11" external cavity wall

4" internal wall

3" trimming joist

firring piece

2" trimmed joists

A   A

3" trimmer joist

herring bone strutting

9" x 9" flue

trimming round doorway

m.s. corbel

2" common or bridging joists

folding wedges

FIG. I.

16"-18" hearth

screeding hearth concrete

flooring

trimmer joist

rough boarding for formwork

FIG. 2. SECTION ON LINE A-A.

Let us assume, for this problem, that each foot run of joist in the floor will carry 98 lbs. of evenly distributed weight, the span of the floor is 10 ft., the breadth of the joists is to be 2 in., and the fibre stress of the timber being used in 900 lbs. per sq. in. The dimensions of the joists would be calculated thus:

$$\frac{Wl}{8} = \frac{fbd^2}{6}$$

$$bd^2 = \frac{Wl}{8} \times \frac{6}{f}$$

$$bd^2 = \frac{980 \times 10 \times 12 \times 6}{8 \times 900} = 98$$

$$bd^2 = 98 = 2 \text{ (breadth)} \times 49 \text{ (depth}^2)$$

$$= 2 \text{ in.} \times 7 \text{ in.}$$

The dimension of the joists will be 2 in. by 7 in.

A chapter on the structural use of timber and its calculations will be included in a future advanced volume.

Typical lay-out. Fig. 1 is the plan of a single upper floor to a domestic house and which involves a fireplace opening, a bay window, and a doorway. If 1 in. tongued-and-grooved flooring is to be used the joists should not be more than 16 in. apart or 18 in. centre to centre.

The first detail to be considered is the fireplace opening. Any timber which is adjacent to a fireplace and which is built into a wall must not be within 9 in. of a flue. Also the hearth to the fireplace must be at least 16 in. wide in front of the opening, and must extend at least 6 in. beyond the opening on each side. It is necessary to stop the joists, immediately in front of the fireplace, at least 16 in. from the opening.

To do this, three joists—a trimmer and two trimming joists— are framed up as seen in the plan of the floor so that the trimming joists run through the full span of the room and rest on the fireplace wall about 3 in. from the chimney breast. The one which is nearest to the flue which comes up from the fireplace in the room below, if built into the wall, must be at least 9 in. from the inside of the flue, but this measurement can be decreased if the end of the joist rests on a metal corbel built into the wall (see Fig. 7). The trimmer joist, which must be at least 16 in. from the fireplace opening is supported at each end by the trimmed joists. Some local authorities insist on 18 in. as being the minimum width of the hearth. The joint used at the intersections is seen in Fig. 4.

As these three joists have to support the weight taken by them-selves and the trimmed joists, they are usually increased in breadth to 3 in. The trimmed joists are supported by the wall at one end and the trimmer joist at the other. They are joined to the trimmer as seen in Fig. 5 which is a splayed housing or by using the common housing joint (Fig. 23, Chapter 3).

The rest of the joists are laid in the two remaining spaces at approximately the same centres as the trimmed joists. All the joists resting on walls must have a bearing of 4 in. When they have all been positioned their ends should be packed up, if necessary, so that all the top surfaces are in one horizontal plane. Lengths of battens should be nailed across their ends to keep them in place until they are built in to the brickwork.

Strutting. So that lateral movements in the joists is cut down to a minimum, herring-bone strutting, similar to that shown in Fig. 6, should be cut and fixed between the joists extending from wall to wall. The spaces between the walls and end joists are occu-pied by folding wedges. The strutting, which is cut to fit against the vertical surfaces of the joists, is fixed with a floor brad near the ends of each piece. The folding wedges at each end, too, must be fixed by nailing.

So that the ends of the floor-boards around the door opening and the splayed portions of the bay window can be fixed, simple trimming, as seen in the drawing (Fig. 1) is necessary. Small pieces of timber cut to fit between the bridging joists and parallel to the splayed walls of the bay window are fixed with nails about 2 in. from the walls. If necessary, small battens can be fixed to the joists to give additional support below the trimming pieces, and these can be about 3 in. in depth and $1\frac{1}{2}$ in. to 2 in. thick.

The trimming around the door opening is a little different. Short pieces should be cut and fitted between the two joists adjacent to the edges of the door opening, and an additional piece cut and fitted between each of these to give a fixing to the board or boards between the jambs of the doorway.

Hearth. To enable the concrete hearth to be formed, a small platform of rough boarding below the hearth opening is fixed on two battens as seen in Fig. 2. If the lower edge of the batten (which is plugged and screwed to the wall) is kept level with the lower edges of the joists this can be used for a fixing for the ceiling material at a later date. It is not necessary to remove the rough formwork for the concrete hearth. The concrete is kept level with the tops of the joists, and the sand and cement screeding laid after the floor-boards have been fixed.

JOINT DETAILS, ETC.

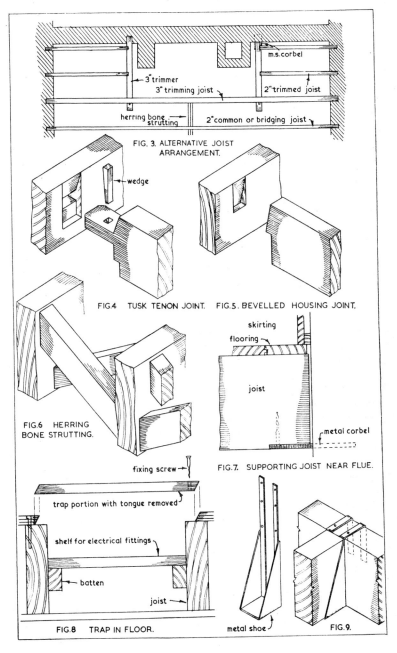

FIG. 3. ALTERNATIVE JOINT ARRANGEMENT.

m.s.corbel

3" trimmer

3" trimming joist

2" trimmed joist

herring bone strutting

2" common or bridging joist

wedge

FIG.4 TUSK TENON JOINT.

FIG.5. BEVELLED HOUSING JOINT.

FIG.6 HERRING BONE. STRUTTING.

skirting

flooring

joist

metal corbel

FIG.7. SUPPORTING JOIST NEAR FLUE.

fixing screw

trap portion with tongue removed

shelf for electrical fittings

batten

joist

FIG.8 TRAP IN FLOOR.

metal shoe

FIG.9.

To allow the ends of the floor-boards which are immediately in front of the chimney breasts to be fixed at their ends, two fixing pieces, dovetailed in section, and approximately 2 in. by 2 in. in dimension, should be pressed into the hearth concrete when the latter has just been poured with their top surfaces level with the top edges of the joists, and extending from the chimney breasts to the trimmer joist.

Alternatively, two short pieces of joist can be fixed between the trimmer and the chimney breast, resting on corbels let into the brickwork at one end (as for the trimming joist), and nailed securely to the trimmer at their other ends.

Alternative joist direction. On occasions, it may be necessary for the joists to run in the opposite direction to that shown in Fig. 1. The method for trimming round a fireplace opening in this second example is shown in Fig. 3. Two short trimmer joists are required with one trimming joist, 16 to 18 in. away from the opening, these forming the space required for the concrete hearth.

Fig. 6 shows a portion of herring-bone strutting which really is two rows of struts throughout the whole width of the floor area. Each piece should be carefully cut to fit between the joists and fixed with floor brads near their ends. Another method for strutting between joists consists of cutting pieces of board up to $1\frac{1}{2}$ in. thick and almost the same depth of the joists so that each piece fits tightly between a pair of the joists. These pieces are nailed in position in a straight line across the width of the floor with folding wedges at each end between the last joist and the wall. This method is not considered as good as the first as they become loose through the joists shrinking.

Metal corbels. Fig. 7 shows how joists, which are not built into a wall, can be supported on a metal corbel.

It is considered very costly these days to get workmen to cut the joints shown in Figs. 4 and 5, and to overcome this expense metal shoes, similar to that shown in Fig. 9, have been produced by manufacturers. These are "hung" on to adjacent joists, and the end of the one to be supported is placed in the shoe as shown.

Electricians require traps in floor-boards so that junction boxes, etc. for their wiring can be fixed and periodically inspected or additional work carried out. All that is required is shown in Fig. 8. A small portion of a floor-board is cut with its ends bevelled. For fixing it in position one or two screws are required. A shelf, made from short ends of floor boards, should be fixed on a couple of battens so that the electrical fittings can be fixed in a workmanlike way.

Double floors. In larger buildings, where the span between the walls is too great for the normal-size timber joist to span, strong

FLOOR DETAILS

**JOISTS OF AN UPPER FLOOR IN POSITION SHOWING STAIR WELL WITH
3 IN. THICK TRIMMER JOIST**

HERRING BONE STRUTTING NAILED BETWEEN JOISTS

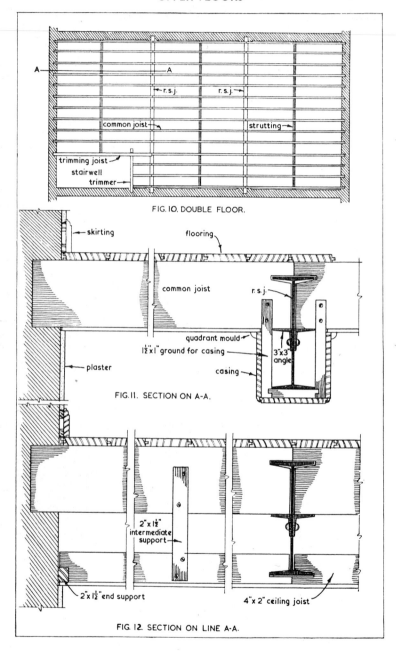

FIG. 10. DOUBLE FLOOR.

FIG. 11. SECTION ON A-A.

FIG. 12. SECTION ON LINE A-A.

intermediate supports must be introduced so that the joists can carry any load for which the floor has been designed. Floors with these intermediate supports are called double floors. Fig. 10 shows the plan of such a floor. The two intermediate British Standard beams to run across the shortest dimension of the floor have been introduced, dividing the longest dimension into three equal parts. Three sets of timber joists run through this direction of the floor, their ends resting on the two metal joists or beams.

Fig. 11 shows a section through a portion of a double floor. The tops of the timber joists should be kept about 2 in. above the top surfaces of the steel joists so that the floor-boards around that area can be fixed securely. The rib of the B.S.B. should be drilled every 3 or 4 ft. or so, so that 3 in. by 3 in. angles can be bolted or riveted to them. The end of the timber joists which rest on the angles are cut so that they fit round, without touching, the flange of the B.S.B's. If they do touch, and shrinkage of the joists takes place, splitting can be the result. The lower edges of the B.S.B's can be hidden by fixing grounds every 2 ft. or 2 ft. 6 in. and a casing fixed to the grounds as seen in Fig. 11.

If it is necessary to have a flat surface for the ceiling, 4 in. by 2 in. joists should be fixed as seen in the lower portion of Fig. 12 to which the ceiling material can be secured. Additional supports for the ceiling members can be provided by nailing or screwing battens to the sides of the floor-joists extending down to the ceiling-joists and nailed or screwed to these. At the wall end of the ceiling, a batten should be plugged to the wall and the ends of the ceiling-joists fitted and nailed to this.

CHAPTER 12   ROOF CONSTRUCTION

ROOFS CAN BE divided into two sections for the purpose of this chapter; single roofs and double roofs. The first are those roofs which have no intermediate transverse supports to the rafters; the second section contains roofs which have their rafters supported once, or more than once, throughout their lengths. Intermediate transverse supports, of course, are commonly called purlins, and we shall see that the rafters of most roofs over a certain span require these supports to prevent their bending under the weight of the roof coverings.

Single roofs. Let us first consider single roofs. Perhaps the commonest of these is the lean-to. This roof consists of a series of rafters with their lower ends resting on a wall plate, which in turn rests on a wall or wooden framing. Their upper ends are securely nailed to a wall piece or wall plate which has been fixed against a vertical surface such as a wall (see Figs. 1 and 2).

Lean-to roof. Fig. 1 shows a vertical section through a lean-to roof. A 6 in. by 1 in. wall piece, to which the top of the rafters are fixed, has been plugged to the wall and a wall plate 4 in. by 3 in., 4 in. by 2 in., or 3 in. by 2 in. in size has been bedded on top of the brick wall and the rafters notched over the top outside corner and securely nailed to it. The ends of the rafters are cut to finish flush with the outside face of the wall so that a fascia or gutter board can be nailed to them along the length of the roof. This type of finish to a roof is called flush eaves, because no part of the roof surface overhangs the outside face of the wall.

Fig. 2 shows an isometric drawing of portion of a lean-to, a wall plate having been placed on the brick corbelling in this instance, to which the tops of the rafters have been notched over and nailed. Ceiling-joists, too, have been added so that a ceiling can be constructed in the building. At the wall end the ceiling joists have been fitted over and nailed to a 2 in. by 2 in. wall piece which has been plugged to the wall. The other ends of the ceiling-joists are nailed to the plate on the wall as have the roof rafters. For additional strength, the joists and the rafters can be nailed together over the wall plate.

In addition to the ceiling-joists giving a fixing to the material to be used for the ceiling, they do also prevent the rafters pushing the

SIMPLE ROOFS

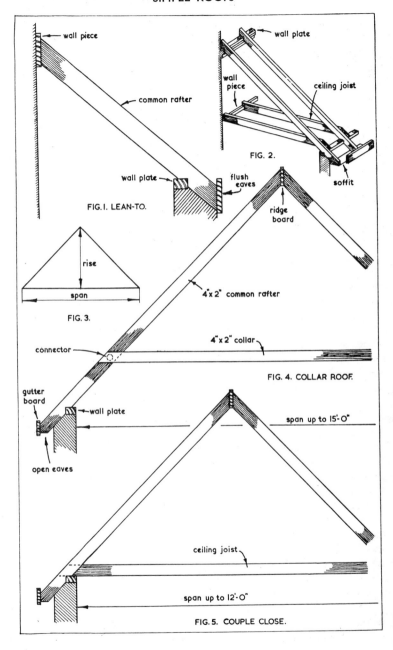

wall piece

common rafter

wall plate

flush eaves

FIG. 1. LEAN-TO.

wall plate

wall piece

ceiling joist

FIG. 2.

soffit

ridge board

rise

span

FIG. 3.

4"x 2" common rafter

4"x 2" collar

connector

FIG. 4. COLLAR ROOF.

gutter board

wall plate

open eaves

span up to 15'-0"

ceiling joist

span up to 12'-0"

FIG. 5. COUPLE CLOSE.

top of the wall outwards. To be sure of this, of course, the ceiling-joists must be securely fixed at both ends. The distance between the centre lines of the rafters should never be greater than 18 in. when roof tiles are to be used for the covering, but this dimension can be adjusted when other materials are used.

Pitch. The covering material, too, will, to a certain extent, control the pitch or angle of slope of the roof. For instance, a tiled roof should never be less than 40 degrees to the horizontal because tiles absorb quite a large amount of water and roofs must be designed to throw off rain-water as quickly as possible before it is absorbed. Slate roofs can be pitched at a much lower angle, to, say, 25 degrees because slate absorbs practically no water at all.

Flat roofs. Flat roofs, which are usually inefficient as far as water is concerned, should have a fall of at least 1 in. in 5 ft. The material used for covering a flat roof, should, of course, be completely unable to absorb water; for instance, sheet lead, copper, asphalt, and good-quality bitumen felt.

Fig. 3 shows the relationship between the rise and the span of a roof. A roof which has a rise which is equal to half the span will have its rafters pitched at an angle of 45 degrees. A roof with a rise of one-third of its span will have the rafters inclined at an angle of approximately 30 degrees.

Sometimes we hear that a roof has a pitch of $\frac{1}{2}$ or $\frac{1}{3}$. This fraction refers to the amount of rise the roof has compared with its span and is sometimes written down thus:

$$\frac{1}{2} \text{ pitch} = \frac{\text{rise}}{\text{span}} = \frac{1}{2} \quad \text{or} \quad \frac{1}{3} \text{ pitch} = \frac{\text{rise}}{\text{span}} = \frac{1}{3}, \text{ etc.}$$

Couple close roof. Fig. 5 is a vertical section through a couple close roof. It is entirely different from a lean-to because it has two sloping surfaces whereas the lean-to has only one. This type of construction can be used for spans up to 12 ft. It is ideal for detached garages for instance. The tops of the 4 in. by 2 in. rafters are nailed to a 6 in. by 1 in. ridge board, and their lower ends notched over and nailed to the plates which are bedded on the top inner edges of the 9-in.-thick walls. Ceiling rafters, 4 in. by 2 in., which overcome any tendency for the feet of the rafters to spread outwards, are nailed to the wall plate and the rafter feet to make a really secure joint. At least four 4 in. nails should be used for the lower joints and two 4 in. nails for the joints at the ridge.

Collar roof. The distance between the unsupported portions of the rafters (between ridge and wall plate) should not be more than 8 ft. If rafters are longer than in the roof shown in Fig. 5 they must

PREFABRICATED ROOF TRUSS READY FOR ERECTION
This is the T.D.A. truss shown in fuller details on page 134.

ROOF IN PROCESS OF ERECTION
The truss shown in the upper photograph is erected in the left-hand bay in the lower illustration.

ROOF CONSTRUCTION

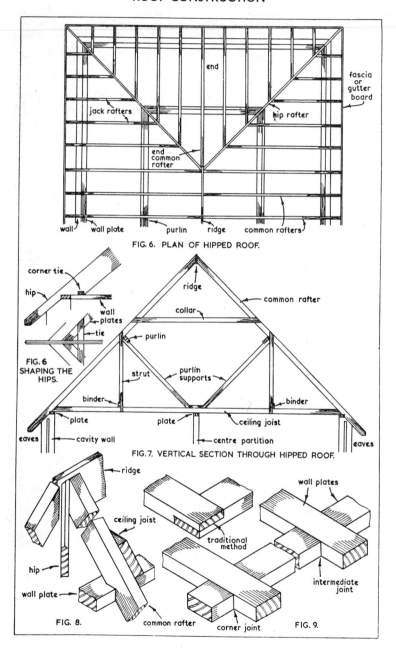

FIG. 6. PLAN OF HIPPED ROOF.

FIG. 6 SHAPING THE HIPS.

FIG. 7. VERTICAL SECTION THROUGH HIPPED ROOF.

FIG. 8.

FIG. 9.

have the intermediate support mentioned at the beginning of the chapter. It is, however, possible to construct a roof up to 15 feet span without introducing purlins by raising the ceiling-joists above wall plate height as seen in Fig. 4. This is a useful type of construction for small-span buildings and provides more headroom between floor and ceiling. The maximum height the ceiling rafters can be raised above the wall plate should be one-third the rise of the roof.

Timber connector. A timber connector, similar to that shown in Fig. 3, Chapter 4, is used to make a really strong joint where each collar, or ceiling joist, is fixed to a rafter. A 5 in. bolt of $\frac{1}{2}$ in. diameter is also used at each joint and two fairly strong washers, preferably square, should also be used, one under the bolt head and the other below the nut. Considerable force is necessary when tightening the nut to make the teeth of the connector penetrate into the wood. If large, square washers are used on the outside surfaces of the timber these will prevent the bolt head and the nut from being forced into the timber themselves.

In Figs. 4 and 5 the feet of the rafters are allowed to extend beyond the outside faces of the walls, and are cut to the shape of a birdsmouth. This finish to a roof is called open eaves, but where a soffit is used, such as in Figs. 2 and 11, the term closed eaves is applied.

Incidentally, the notched joint where the rafter is jointed to the wall plate (Fig. 2) is often called a birdsmouthed joint, but this is a wrong term. The birdsmouth is at the foot end of the rafter when it is shaped as in Figs. 4 and 5.

Wide span. Most craftsmen who have roofing as part of their everyday life spend most of their time on roofs to domestic houses, and these usually have rafters of such a length as to require the intermediate supports called purlins. These roofs can be either gable roofs or hipped roofs, but both kinds have to be constructed and supported similarly. The gable roof has two parallel surfaces, equally pitched usually, and all the rafters are called common rafters.

Hipped roofs. The hipped roof has three or more surfaces, and, when the plan is regularly shaped, the end surfaces are at right angles to the roof's longer sides. Fig. 6 is part of the plan of a hipped roof and shows that the roof has three kinds of rafters; common rafters, which extend between the eaves and ridge; hip rafters, along which the adjacent roof surfaces intersect, and which extend also from the eaves to the ridge; and lastly, jack rafters, which run from the eaves up to the hip rafters.

Fig. 7 shows how this type, and the gable roof are supported in the traditional manner. First of all the wall plates must be bedded on the tops of the walls, and should be checked for squareness. The joints used at the intersections of wall plates are seen in Fig. 9. A much stronger joint is obtained if cross-halvings are used in preference to the traditional halving joint.

The positions of the rafters should be marked on the top surfaces of the wall plates. The ceiling-joists can be positioned and nailed in place alongside the marks on the wall plates. Three or four pairs of common rafters should be pitched and secured to the ridge and wall plates with 4 in. nails. A temporary brace can be nailed across their surfaces to keep them in vertical alignment, and to stop them from being blown over by the wind.

The hip rafters and end common rafters should now be cut and fixed, the whole thus becomes a fairly rigid structure.

Purlins. It is now that the 6 in. by 2 in. purlins and their supports should be cut and fixed, because the later these are left the more difficult they are to get into position. The purlins should be positioned half-way between the ridge and wall plates with one of their narrow surfaces supporting the underneath surfaces of the common and jack rafters. The ends of the purlins should be cut to the correct bevels so that they fit up against the wide surfaces of the hip rafters, and they should be securely nailed in that position.

Before the fixing of the purlins is carried out a couple of 4 in. by 2 in. purlin supports should be fixed under each surface of the roof so that the purlin will be supported immediately it is offered up for fixing. These 4 in. by 2 in. or 3 in. by 3 in. supports are nailed to the common rafters at the top. Their lower ends rest on top of a ceiling-joist, each pair being prevented from sliding by a piece of 2 in. by 2 in. timber cut between them and nailed to the top edge of the joist. It is, of course, advisable to have the feet of the supports as near to an internal partition wall as possible. The tops of the supports should be housed about $\frac{1}{2}$ in. to receive the purlin.

Struts, etc. When the purlins have been fixed, the remainder of the supports and 3 in. by 3 in. struts should be fixed, the distance apart being about every six rafters along the lengths of the roof surfaces. Collars, of 4 in. by 2 in. stuff, too, should be fixed above the supports as seen in Fig. 7. To prevent the ceiling-joists from bending under the weight of the material used for the ceilings, 4 in. by 2 in. binders should be fixed to the purlin struts along the lengths of the roof at right angles to the joists. All joists are securely nailed to the binders. The struts should be housed out to a depth of, say, $\frac{1}{2}$ in. to receive the binders so that there can be no danger of the nailed joint coming apart. The remaining common rafters and jack rafters can now be fixed. Fig. 8 shows the main joints between the various parts of a roof.

INTERSECTION OF WALL PLATES WITH CROSS-HALVINGS

RAFTER NOTCHED OVER WALL PLATE. THE MEMBER BOLTED TO THE RAFTER IS THE BINDER WHICH CAN BE SEEN AT THE BOTTOM OF THE T.D.A. HALF TRUSS IN FIG. 21. THE FOURTH MEMBER IS A CEILING JOIST

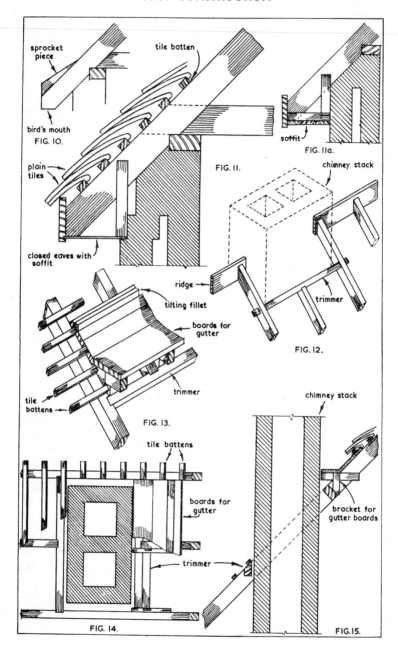

sprocket piece

tile batten

bird's mouth
FIG. 10.

plain tiles

closed eaves with soffit

FIG. 11.

soffit

FIG. 11a.

chimney stack

ridge

tilting fillet

boards for gutter

trimmer

FIG. 12.

tile battens

trimmer

FIG. 13.

tile battens

boards for gutter

trimmer

FIG. 14.

chimney stack

bracket for gutter boards

FIG. 15.

Eaves. Figs. 10 and 11 show details around the eaves of a roof. Often it is desired to have a slight upwards sweep of the roof surfaces at the eaves, and this can be obtained by nailing shaped sprocket pieces on the tops of all the rafters as in Fig. 10. Fig. 11 shows to a larger scale, details of what are called "closed eaves", the feet of the rafters being allowed to overhang a distance beyond the surface of the wall. If the feet of the rafters are not to be seen a soffit has to be fixed between the fascia or gutter board and the wall.

If, as is often the case, asbestos cement sheets are used for the soffit, and the wall is constructed as in Fig. 11, the sheeting can be screwed to the feet of the rafters and also rest on the small ledge on the wall. A small batten can be nailed to the rafters every 3 ft. or so to hold the soffit down securely on the ledge. If on the other hand, the soffit is to be matching or similar material, the fixings for the soffit should be similar to that shown in Fig. 11a.

Chimney stack arrangements. Where chimney stacks pass through the roof a certain amount of additional work is involved in the trimming of the woodwork to form the openings for the chimneys. If the stack passes through the centre of the roof the trimming is carried out as seen in Fig. 12. A trimmer is introduced on each side of the ridge so that certain rafters can be cut short to form the opening. The trimmer can be equal in size to the rafters, and each joint can be similar to that shown in Fig. 23, Chapter 3, or mortice and tenon joints as seen in Fig. 12 can be used.

Some chimney stacks pass through roof surfaces at some distance from the ridge, and a vertical section through one of these is seen in Fig. 15. Two trimmers on the one surface are required for this type of stack, the plan (Fig. 14) showing these clearly. When possible, it is an advantage to keep the lower trimmer in vertical position with its top surface bevelled so that the tile batten adjacent to the stack can be kept near to the brickwork, thus keeping the top of the tile close to the brickwork. The top trimmer can be fixed with its top surface parallel with the roof surface as seen in the drawing.

Preparation also has to be made for the gutter the plumber is required to fix behind the stack. This involves making three or four brackets on to which the gutter boarding can be nailed (see Figs. 13 and 15). The minimum width and depth of the gutter is 6 in. and, in addition to boards being nailed on top of the brackets, they also have to be nailed up the slope of the rafters so that a tilting fillet for the lowest tiles can be fixed at least 6 in. above the gutter bottom, measured vertically. Figs. 13, 14, and 15 give details of this type of construction.

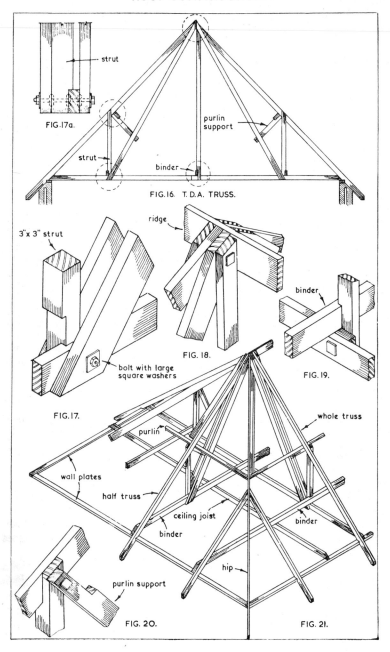

FIG.17a.

strut

strut

purlin support

binder

FIG.16. T.D.A. TRUSS.

ridge

3"x 3" strut

bolt with large square washers

FIG. 18.

binder

FIG. 19.

FIG.17.

purlin

whole truss

wall plates

half truss

ceiling joist

binder

binder

purlin support

hip

FIG. 20.

FIG. 21.

T.D.A. roof truss. Perhaps the best advancement in house construction since the war has been the development of really efficient roof trusses by the association which was at that time called The Timber Development Association. They revolutionised the methods for constructing timber roofs of all descriptions. That shown in Fig. 16 was developed for the small domestic type of building.

It consists of trusses made from fairly small section material, 3 in. by 2 in., 4 in. by 1½ in. or 4 in. by 2 in., bolted at the joints. These trusses are fixed at intervals of six rafters or so along the lengths of the roof. Between each pair of timbers, at the joints, a double-toothed timber connector is used making it a most rigid frame capable of carrying fairly large loads. In addition to this, all the weight of the roof is thrown on to the external walls leaving the internal walls non-load-bearing. If necessary, the trusses can be made up in two halves, for easy handling, the ceiling-joist being in two pieces. When in position the two halves can be secured by fish-plates. Half-trusses are used at the ends of a hipped roof. One of these is shown in position in the isometric drawing (Fig. 21).

The procedure for constructing the traditional roof can be employed for this modern type. Instead of erecting three or four pairs of common rafters, the T.D.A. trusses, as they are called, will be the first timbers to be fixed after the ceiling joists. Figs. 17, 18, 19, and 20 show details of the joints used for these trusses.

Fig. 17a is a section through the timbers shown in Fig. 17. For this joint three connectors would be required.

ROOF GEOMETRY

Now let us consider the geometry involved in obtaining the lengths and bevels for the various parts of a roof. Fig. 22 is a vertical section through a roof pitched at 45 degrees, and Fig. 23 is the plan of one end of the roof which is hipped.

Common rafters. First the common rafters. The two dimensions required to obtain the length of the common rafters of a roof are the horizontal distance between the top outside corners of the wall plates a–b, and the pitch or angle of the slope of the roof. Sometimes the information is given as the distance the roof rises for every foot run, measured in a horizontal direction.

The length of the common rafter is a–e, which is the length of a line from the top outside corner of the wall plate a to point e which is on the centre line of the ridge board. The distance from a to the lower end of the rafter has to be added on to the length obtained, and this addition depends on the amount of overhang, if any, of the

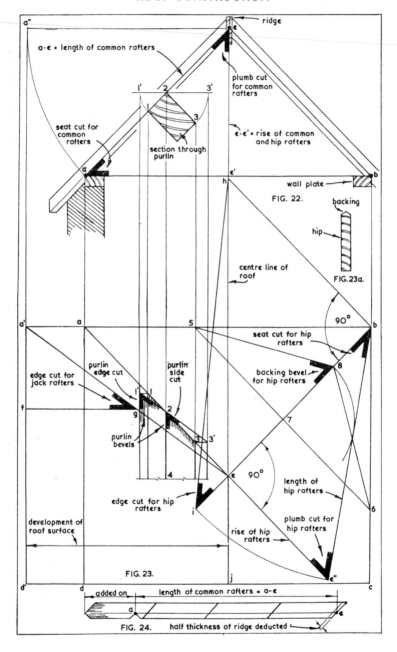

a″

ridge

a-e = length of common rafters

plumb cut for common rafters

seat cut for common rafters

e-e′ = rise of common and hip rafters

section through purlin

FIG. 22.

backing

hip

centre line of roof

FIG. 23a.

seat cut for hip rafters

90°

edge cut for jack rafters

purlin edge cut

purlin side cut

backing bevel for hip rafters

purlin bevels

length of hip rafters

90°

edge cut for hip rafters

development of roof surface

rise of hip rafters

plumb cut for hip rafters

FIG. 23.

added on

length of common rafters = a-e

FIG. 24.   half thickness of ridge deducted

eaves. Half the thickness of the ridge board will have to be deducted from the length at the top end of the rafter.

Fig. 24 shows how the first common rafter is marked out. First the depth of the notching is decided upon. This is usually one-third the depth of the rafters, and a line this distance from the lower edge is marked through the whole length of the piece of timber being marked. A sliding bevel is set up to the plumb cut, and this bevel should be marked on the side of the timber at the top end. From point e, and on the line marked along the length of the timber, the distance a–e should be marked to give point a. The plumb and seat bevels should be applied to pass through point a to give the shape of the notching.

The additional length required at the foot of the rafter should then be calculated and marked from a, the plumb and seat bevels again being applied to the end to obtain the birdsmouth. Half the thickness of the ridge board (measured at right angles to the bevel) should be deducted from point e. When the rafter has been cut to the correct shape it can be used as a templet to mark all the other common rafters.

Jack rafters. The number of jack rafters to each end of the roof should be calculated, and the common rafter templet divided into a number of equal parts to give the lengths of the jack rafters. The templet (Fig. 24), shows that three jack rafters are required at the ends of each surface.

Hip rafters. The plan of the roof is shown in Fig. 23. The lines a–b, b–c, and a–d, represent the top outside corners of the wall plates, and line e–j represents the centre line of the ridge board. To obtain the length of hip b–e, construct a right angle at e, and make e–e'' equal to the rise of the common rafter e–e'. The rise of the hip rafter is always equal to the rise of the common rafter. The line b–e'' represents the length of the hip rafter b–e, and, as the roof is regular in its shape, the other hips are also equal in length to it.

The plumb and seat bevels are shown at e'' and b respectively. Another bevel in addition to the plumb cut is required to apply to the top of the hip rafter so that the cut surface will bed up against the ridge board. This bevel is called the hip edge cut. To develop this, construct a right angle at b to intersect with the extended centre line of the ridge in h. With centre b and radius b–e'' describe the arc to give point i on the extended plan of the hip rafter b–e. Join h to i with a straight line to give the bevel required.

Remember, half the thickness of the ridge board, measured at right angles to the hip edge bevel, must be deducted from the length when cutting the bevels.

It may also be necessary to apply a backing bevel to the top edges of the hip rafters to give a seating to boarding or tile battens. This backing bevel is equal to half of the dihedral angle set up by two intersecting roof surfaces. To develop the dihedral angle, draw line 5–7–6 at right angles to hip b–e. This can be placed anywhere between b and e. With centre on point 7, and with compasses opened so that the pencil just touches b–e'' describe an arc to give point 8 on b–e. Join 5 to 8 and 6 to 8 with straight lines to develop the dihedral angle 5–8–6. The backing bevel is 5–8–7, and is applied on both sides of the centre line on the top edge of each hip rafter as shown in Fig. 23a.

Jack rafter bevels. The jack rafters also require a second bevel to be applied to their top ends so that they fit up against the hip rafters. This bevel is called the jack rafter edge bevel, and is ascertained by first developing one of the surfaces of the roof. With centre e in the section, and with radius e–a describe an arc to give point a'' on a horizontal line drawn from e. Drop a vertical line from a'' to give points a' and d' on horizontal lines brought out from a and d respectively; e–a'–d'–j is the roof surface e–a–d–j developed. Any line drawn at right angles to a'–d' to intersect with a'–e will give the jack rafters edge cut.

Purlins. So that the ends of the purlins will fit up against the sides of the hip rafters, two more bevels must be developed. These are the purlin side cut and the purlin edge cut. To obtain these bevels one of the narrow surfaces and one of the wide surfaces of a purlin must be developed.

Draw a section of one of the purlins (any size will do) at right angles to the common rafter as shown in Fig. 22. Draw a horizontal line through point 2.

Project points 1, 2 and 3 vertically downwards to give these points on the plan of the hip rafter a–e. With compass point in 2 and radius 2–1 and 2–3 in turn, describe arcs to give points 1' and 3' on the horizontal line passing through point 2. Drop vertical lines from 1' and 3' downwards to intersect with horizontal lines brought out from 1 and 3 in the plan to give points 1' and 3'. Join 2 to 1' and 2 to 3' to obtain the two bevels required. These bevels are applied to the two ends of each purlin. The cut surfaces should fit up against the wide surfaces of the hip rafters.

Flat roofs. Flat roofs are comparatively easy to construct compared with pitched roofs. Fig. 25 is the plan of a flat roof with an eaves overhang of 6 in. So that water will flow from the surface of a flat roof it must be given a minimum amount of slope. This slope should be approximately 1 in. in every 5 ft. In other

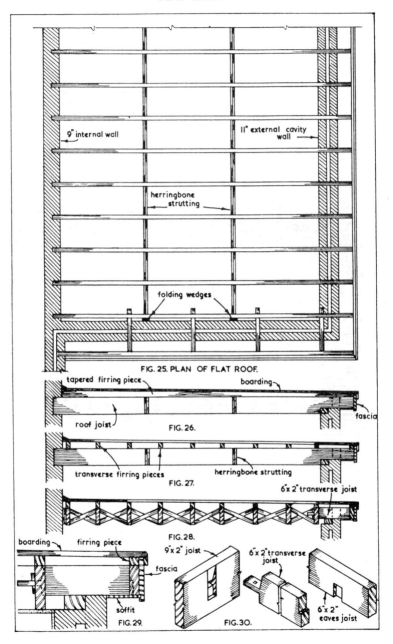

FIG. 25. PLAN OF FLAT ROOF.

9" internal wall

11" external cavity wall

herringbone strutting

folding wedges

tapered firring piece

boarding

roof joist

fascia

FIG. 26.

transverse firring pieces

herringbone strutting

FIG. 27.

6"x 2" transverse joist

FIG. 28.

boarding

firring piece

fascia

soffit

FIG. 29.

9"x 2" joist

6"x 2" transverse joist

6"x 2" eaves joist

FIG. 30.

WALL PLATES IN POSITION READY FOR ROOFING

TILE BATTENS NAILED DOWN ON TO RAFTERS

words, a flat roof 10 ft. wide should have a fall towards the gutter of at least 2 in.

Several methods are used to obtain the fall to a flat roof, three of which are shown in Figs. 26, 27, and 28. In each case the joists of the roof have been laid horizontally to give a level ceiling in the space below. Fig. 25 shows that the slope has been obtained by nailing on top of the joists tapered firring pieces, the thinnest parts being about $1\frac{1}{2}$ in. in thickness. Boards have to be nailed on top of the firring pieces to obtain the surface on which the material to be used for waterproofing the roof is placed.

The second example (Fig. 27) shows the firring pieces nailed at right angles to the joists. In this case the firring pieces are not tapered but are all of different widths.

The third example is similar. The firring pieces are nailed on the tops of the joists, but the fall of the roof is at right angles to the run of the joists. Consequently, the firring pieces are not parallel but all different in width as in Fig. 27. To stop any possible lateral movement of the timbers, herring-bone strutting has been introduced between the joists. These rows of strutting should not be more than 5 ft. apart. Folding wedges should be inserted at both ends of the strutting between the last and first joists and the brickwork.

Eaves. Two eaves' details are given to show how the overhang is constructed. In Figs. 26 and 27 the ends of the joists rest on a wall plate, and are reduced in width for the eaves, the gutter board and soffit completing the work.

In Fig. 28 the eaves have been constructed by introducing short, transverse, 6 in. by 2 in. joists, because the main joists run in the opposite direction to the first two examples. Larger details of this are shown in Fig. 29, and the joints used are seen in Fig. 30.

Also see chapter 23.

CHAPTER 13   STUD AND GLAZED PARTITIONS

PARTITIONS ARE A means of dividing a large space in a house, or any other type of building into smaller compartments such as bedrooms, offices, stock rooms, etc.

Stud partitions.   Where no windows are required in the dividing partition the common stud partition (Fig. 1) is often used.   This is simple to construct, the only item which requires any special consideration being the doorway.   The partition is constructed of 4 in. by 2 in. or 3 in. by 2 in. members throughout, and consists of a sill which is securely nailed or screwed to the floor, a head similarly fixed to the ceiling-joists, two vertical side studs, plugged to the walls, and intermediate studs.

If the doorway can be positioned to suit the size of sheeting to be used for facing the partition, so much the better.   4-ft.-wide sheets are commonly used for this kind of work, and the broken lines in Fig. 1 show the positions of the covering sheets.   The centres of the vertical studs are 2 ft. apart, measuring from the surface of the wall farthest from the door opening, if any.   The door opening consists of two vertical studs and a head, the joints at the head being similar to the mortise and tenon joint, Fig. 30, Chapter 3.   Those at the feet of the studs where they intersect with the sill are shown in Fig. 2.

The width of the door opening will, of course, depend on the width of the door, and the thickness of the door linings must also be taken into consideration (see Fig. 3).   This must also be kept in mind when the door head is being positioned.   Noggings or stiffening pieces are horizontal pieces of timber similar to the studs, and are fixed at regular intervals on the height of the partition.   These not only give lateral support to the structure but also additional fixing for the surface coverings.   All the pieces forming the partition are securely nailed together with 3 in. or 4 in. wire nails.   Architrave fixing strips, equal in thickness to the covering sheets, are nailed around the door opening, and these architraves should be wide enough to cover the joints between the strips and the sheeting (Fig. 3).

There are many ways of soundproofing partitions, and with ½ in. insulating board covering each surface of the partition, a fair amount

SIMPLE PARTITION

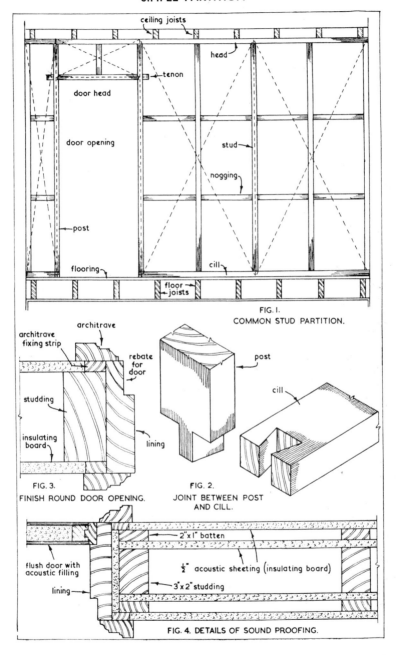

FIG. I.
COMMON STUD PARTITION.

FIG. 3.
FINISH ROUND DOOR OPENING.

FIG. 2.
JOINT BETWEEN POST
AND CILL.

FIG. 4. DETAILS OF SOUND PROOFING.

STUD AND GLAZED PARTITIONS

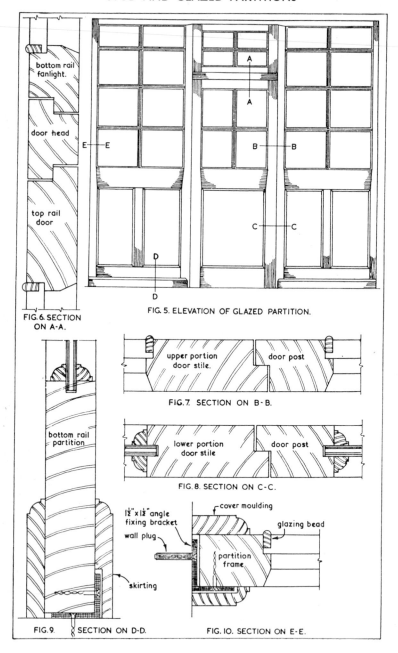

bottom rail fanlight.

door head

E — E

top rail door

FIG. 6. SECTION ON A-A.

A

A

B — B

C — C

D

D

FIG. 5. ELEVATION OF GLAZED PARTITION.

upper portion door stile.

door post

FIG. 7. SECTION ON B-B.

bottom rail partition

lower portion door stile

door post

FIG. 8. SECTION ON C-C.

1½" x 1½" angle fixing bracket

wall plug

skirting

cover moulding

glazing bead

partition frame

FIG. 9. SECTION ON D-D.

FIG. 10. SECTION ON E-E.

of soundproofing will have been obtained. If this is not sufficient, the method shown in Fig. 4 could be adopted. This involves constructing a partition similar to that in Fig. 1 and fixing 2 in. by 1 in. battens on top of the sheeting and immediately in front of the studs, nogging pieces, sill, and head. Each face is covered with another thickness of insulating board. The insulating board should also be fixed immediately behind the door linings. This construction requires wider linings, and in a case like this both edges could be rebated as shown to break up the wide surface between the door and architrave. The door, too, should be of a construction which will give adequate soundproofing.

Glazed partitions. In office buildings glazed partitions are often required so that adequate light and ventilation can be provided to the space behind the partition. Light, flush partitions which can be fixed fairly easily are also a requirement, and the glazed partition and its details shown in Figs. 5–10 is an example of this kind of construction.

The partition framing and the door are both equal in thickness which necessitates the edges of the door opening being rebated as well as three edges of the door, see Figs. 6, 7, and 8. For ventilation purposes, a fanlight has been provided above the doorway, and the four edges of the opening and the four edges of the fanlight will also have to be rebated similarly to the doorway (see Fig. 6).

The framework of the partition has been designed similarly to that of the door so that the horizontal members will follow through the width of the partition with unbroken lines. Panels of $\frac{3}{8}$-in.-thick plywood are included below the middle rail, with ovolo planted mouldings fixed around the edges of the frames on both sides. Glazing beads are provided for all the glass rebates on the inside of the partition and door (see Figs. 6, 7, and 10).

The partition is fixed by means of $1\frac{1}{2}$ in. by $1\frac{1}{2}$ in. angle brackets approximately 3 in. in length. These are drilled and countersunk to receive number twelve screws. The brackets are lined up and screwed to the floor and ceiling about 3 ft. apart, and also in vertical alignment of the faces of the two walls, again about 3 ft. apart. The partition is placed in position so that it butts up against the brackets, and the fixing screws are driven home.

The brackets and the spaces between the framing and floor, ceiling and walls are hidden by fixing skirtings along the floor, and cover mouldings along the ceiling line and walls on both sides as seen in Figs. 9 and 10. The skirtings and mouldings will have to be recessed out at the bracket positions.

CHAPTER 14   WINDOWS

WINDOWS, WHICH GIVE ventilation and light to a building, can be divided into three types. The first group includes those casement windows which are hinged and open either inwards or outwards or are fixed; the second group are those which are pivoted so that when opened the surfaces of the glass are inclined to the vertical surfaces of the building; and lastly are those which slide in a sideways, upwards, or downwards direction. In addition to giving light and ventilation it is important to see that the windows and their frames are designed and constructed in such a way as to exclude all moisture and draughts from the inside of the building.

Casement windows. Casement windows, Figs. 1, 2, and 3, are those which fall into the first group. They can either be fixed so that they do not open or they can be hinged to open outwards or, as in the case of fanlights, if required they can open into the building.

Fig. 1 shows a vertical section through a casement and frame. The frame, which comprises a head, sill, and two posts is made from 4 in. by 3 in. material. A rebate, $\frac{1}{2}$ in. deep and $1\frac{1}{2}$ in. wide, is worked on the external edges of the frame to receive a $1\frac{3}{8}$-in.-thick casement. The internal edges of the frame are moulded. The rebate of the sill, which should be made from a hardwood such as oak, should be bevelled as the drawing shows so that any water settling on the top surface will be thrown to the outside of the building. A groove, $\frac{1}{4}$ in. by $\frac{1}{4}$ in. should be worked on the lower surface of the sill near to the front edge so that water rolling down the vertical surface will be prevented from penetrating into the brickwork below. This groove is called a drip. Cement grooves, on the four outside edges of the frame, also help to prevent the penetration of water into the building.

The frame is placed in position when the brickwork has reached sill height and it is fixed by screwing to the sides metal cramps which are ragged at their ends as seen in Fig. 15, Chapter 4. They are built into the brickwork as it is carried up and the cement grooves are filled with mortar as the work progresses.

The casement is $1\frac{3}{8}$ in. thick, the two stiles and top rail being $1\frac{1}{4}$ in. wide and the bottom rail, which should be a little wider than the other parts, is about $2\frac{1}{2}$ in. wide. Rebates, $\frac{1}{4}$ in. deep and $\frac{5}{8}$ in.

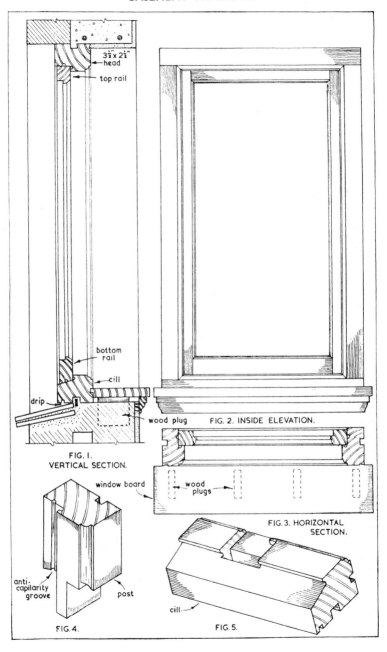

$3\frac{1}{2}'' \times 2\frac{1}{2}''$
head

top rail

bottom rail

cill

drip

wood plug

FIG. 1.
VERTICAL SECTION.

FIG. 2. INSIDE ELEVATION.

window board

wood plugs

FIG. 3. HORIZONTAL SECTION.

anti-capilarity groove

post

cill

FIG. 4.

FIG. 5.

WINDOWS

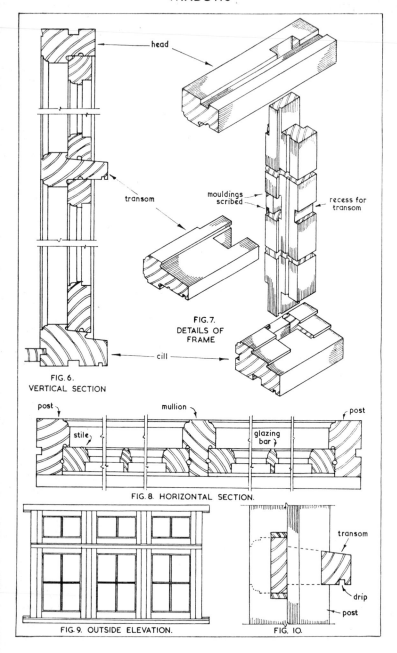

FIG. 6.
VERTICAL SECTION

FIG. 7.
DETAILS OF
FRAME

head

transom

mouldings
scribed

recess for
transom

cill

FIG. 8. HORIZONTAL SECTION.

post

mullion

stile

glazing
bar

post

FIG. 9. OUTSIDE ELEVATION.

transom

drip

post

FIG. 10.

wide, approximately, are worked on the external edges of the casement to receive the glass, and a moulding similar to that on the frame is worked on the internal edges. Joints similar to that shown in Fig. 29, Chapter 3 should be used for making the casement.

The window-board, which should be fixed before the plastering is done, is secured in position by nailing to the wood plugs which are built into the joints of the brickwork. To avoid unsightly shrinkage gaps a groove should be worked on the inside surface of the sill to receive the tongue on the edge of the window board. A wood ground, equal in thickness to the plaster, should also be fixed along the top edge of the brickwork below the window-board so that the apron moulding can be fixed to it when the plastering has been completed.

Capillary grooves should be worked on the two surfaces of the frame rebates to prevent the penetration of water by capillarity between the surfaces of the frame and casement. The exception is on the wide surface of the rebate on the sill. Water would collect in this groove so the second groove around the sill is worked on the lower edge of the bottom rail of the casement. Fig. 3 shows the positions of the wood plugs below the window-board. Figs. 4 and 5 show the mortise and tenon joints for the frame. The mouldings are scribed at their intersection.

Casement with fanlights. Fig. 9 is the outside elevation of a casement frame with fanlights. The vertical members between the posts are mullions, and the horizontal member between head and sill is a transom. Fig. 6, which is a vertical section through the frame shows that the sill and transom are both wider than the other parts of the frame, and that the top rebate of the transom has to be bevelled in the same way as that of the sill. The latter, too, has been double rebated as an extra precaution against water getting into the building. Capillary grooves are again worked on the rebate surfaces as before. Fig. 8 is the horizontal section through the frame. Fig. 7 shows some of the joints used in the frame, and special note should be made of the recess for the transom on the outside edge of the post. This should be approximately $\frac{1}{4}$ in. in depth.

Fig. 10 is an enlarged view of the post around the position of the transom. It shows that the front edge of the transom is housed into the post, and also gives the position of the mortise and tenon joint.

Bay window with casements. Casement windows are, more often than not, used for bay windows and part of the plan of one of these is shown in Fig. 11. This shows that the two side lights are inclined to the centre portion at an angle of 45 degrees. The sill and head of the frame are of three pieces each, and are connected together by means of a handrail bolt and two dowels as seen in Fig. 13.

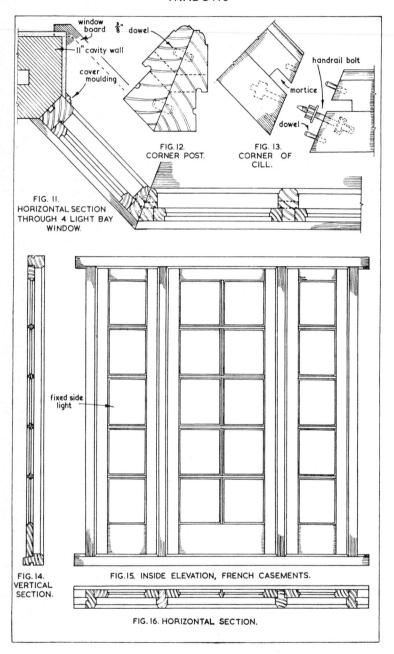

window board ⅜" dowel

11" cavity wall

cover moulding

FIG. 12.
CORNER POST.

handrail bolt

mortice

dowel

FIG. 13.
CORNER OF
CILL.

FIG. 11.
HORIZONTAL SECTION
THROUGH 4 LIGHT BAY
WINDOW.

fixed side light

FIG. 14.
VERTICAL
SECTION.

FIG. 15. INSIDE ELEVATION, FRENCH CASEMENTS.

FIG. 16. HORIZONTAL SECTION.

The corner posts are made up from two pieces of timber, glued and dowelled together, or can be prepared from one piece.

Figs. 12 and 13 also indicate the type and position of the mortice and tenon joints which are used at the intersections with the corner posts with the head and sill. The joints should be glued with a synthetic adhesive and dowelled. The window-frames are fixed in a similar way to ordinary casement windows using wall ties screwed to the sides of the frame. The fanlights can be either hinged at the top to open outwards or at the bottom to open into the building.

French Casements not only give light and ventilation to a building but access also. These are usually found at the rear of a building, and open out onto a lawn or terrace.

The french casement in Fig. 15, has a fully glazed door in the centre with a fully glazed fixed light on each side of the doorway. The door-opening should be at least 6 ft. 6 in. high and 2 ft. 6 in. wide. If the frame is of sufficient height a transom can be incorporated with opening or fixed fanlights above. When required to open, they should be hinged at their tops to open outwards or at their lower edges if it is required to open them into the room. The sill and transom, if incorporated into the frame, should have their top rebate well bevelled to throw off any rain or moisture which may settle on them. Figs. 14 and 16 are the vertical and horizontal sections respectively.

Pivot-hung sashes. Pivot-hung sashes are becoming popular once again and these can either be pivoted at their edges, as in the case of Figs. 17 and 18, or they can be pivoted about the top and bottom rails. Fig. 17 shows a vertical section through a casement and frame with the pivots fixed half-way down the stiles. In this case it will be seen that the casement is placed centrally in the frame with stop beads on each side. The head and sill are rebated with the wider surface in each case well bevelled. Capillary grooves are worked in the rebates as before.

To obtain the positions of the bevelled cuts in the stops running down the sides of the frame the casement should first be drawn in the open position (see Fig. 18). The casement in this drawing has been drawn at an angle of 90 degrees to the frame but this can be adjusted to suit any special requirements.

To obtain the cuts draw the line a–b and make angle a–b–c a right angle. This will give the cuts to the stops below the casement. To obtain the cuts above the casement make angle b–a–d 90 degrees and draw a line parallel to a–d at least 1 in. away. This is the position of the top cuts. To complete the drawing, with compass point in the centre of the pivot, and opened to touch the edges of the cuts to the stops on the frame, arcs can be drawn to give the positions of the cuts to the stops on the casement. The reason for keeping the cuts to the stops above the casement well away from point *a* is to allow the easy removal of the casement if necessary.

The pivots (Fig. 20) are housed into the inside surfaces of the frame sides, and the socket plate is housed into the sides of the stiles of the casement. So that the casement can be placed into, and removed from, its position, a groove has to be made to the inside edge of the stile, as seen in Fig. 19, and the groove is worked in the edge of the stop along to its end. To remove the casement from the frame the bottom ends of the two stiles are firmly grasped with the hands and the casement is pushed forward and upwards so that the pivots slide along the grooves made in the stiles and stops. The reverse movements are made to replace the sash. Fig. 21 shows the joints suitable for the frame.

Linings. Linings to the openings for doors and windows are often required, and when the thickness of the wall is such that it will prevent adequate light from entering the building, splayed linings are often employed similar to those shown in Figs. 22 and 23. In this case the side or jamb linings are splayed out at an angle of 45 degrees to the face of the wall, and the head lining is horizontal. Grooves are made in the surfaces of the frame to receive the tongues on the edges of the linings. Grounds are fixed to the wall near the inside edges of the linings so that they and the architraves can be secured in position. The grounds should be fixed before the plastering is commenced.

Sliding sashes. Double-hung, or vertical sliding sashes, too, are being used much more in recent years than they have for a long time. Usually they consist of two sashes in a built-up frame, and these are able to slide upwards or downwards in a groove.

Fig. 25 is the inside elevation of one of these window-frames with the two sashes in the closed position. Fig. 24 is a vertical section through the sashes and frame. The main portion of the frame consists of four pieces as in any other frame, and these are a sill, a head, and two pulley stiles (see Fig. 31). Spaces behind the two pulley stiles are required for the weights which are used to counterbalance the sashes. Without such counterbalancing the top sash would not remain in its closed position but would fall down to the sill, and the bottom sash could not be left in the open position.

The spaces or boxes for the weights are formed by fixing inside and outside linings to the pulley stile by means of tongued and grooved joints. This is shown in Fig. 26 which is a horizontal section through the frame and sashes. The fourth side of the box,

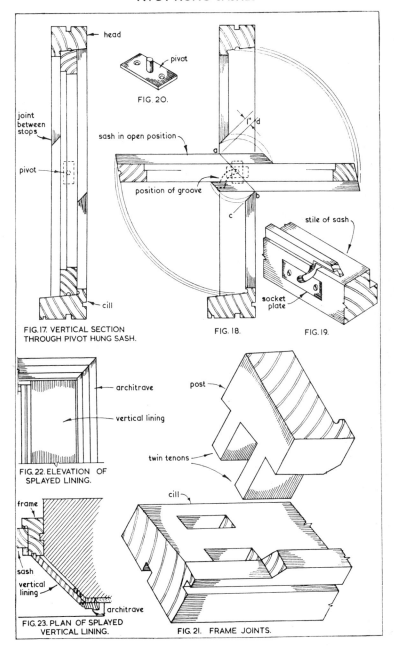

head

pivot

FIG. 20.

joint
between
stops

pivot

cill

sash in open position

position of groove

1" d

a

b

c

stile of sash

socket
plate

FIG. 17. VERTICAL SECTION
THROUGH PIVOT HUNG SASH.

FIG. 18.

FIG. 19.

architrave

vertical lining

FIG. 22. ELEVATION OF
SPLAYED LINING.

frame

sash

vertical
lining

architrave

FIG. 23. PLAN OF SPLAYED
VERTICAL LINING.

post

twin tenons

cill

FIG. 21. FRAME JOINTS.

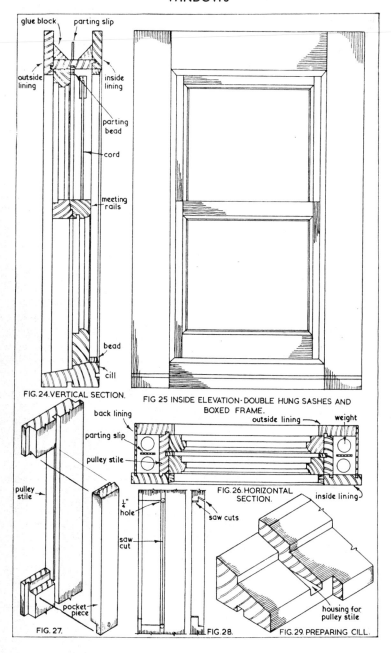

glue block · parting slip

outside lining

inside lining

parting bead

cord

meeting rails

bead

cill

FIG. 24. VERTICAL SECTION.

FIG 25 INSIDE ELEVATION · DOUBLE HUNG SASHES AND BOXED FRAME.

back lining

parting slip

pulley stile

outside lining — weight

FIG. 26. HORIZONTAL SECTION.

inside lining

pulley stile

$\frac{1}{4}$" hole

saw cuts

saw cut

pocket piece

housing for pulley stile

FIG. 27.

FIG. 28.

FIG. 29. PREPARING CILL.

144

the back lining, is formed by fixing a piece of $\frac{1}{4}$ in. plywood or other thin piece of timber to the back edge of the outside lining as seen in the drawing. The spaces inside the boxes so formed must be large enough to take two weights each. These are kept apart by hanging a parting slip from each end of the head (see Fig. 31). Inside and outside horizontal linings, similar to those used to form the weight boxes, are used across the top of the frame and are fixed by means of tongued and grooved joints to the head.

The sashes are kept about $\frac{3}{8}$ in. apart by means of parting beads which are fitted into grooves worked on the surfaces of the pulley stiles and head. The inside edges of the external linings are allowed to project beyond the surfaces of the pulley stile and head to a distance of approximately $\frac{5}{8}$ in. to form the edge of the groove or housing for the top sash, but the edge of the housing for the bottom sash is formed by a loose bead which is nailed or screwed to the edge of the inside linings. This loose bead when removed will allow the sashes to be taken out for repair if necessary. The parting beads, too, are removable for the same purpose.

The sill has to be prepared at each end in a similar way to that shown in Fig. 29. Housings, equal to the thickness of the pulley stiles, have to be made near the ends of the sill, and care should be taken to make the edges of the housing truly vertical. The pulley stiles should, when fitted into the housings, be a tight fit. They are secured by gluing, using preferably a synthetic resin adhesive, and a couple of nails through the back surfaces of the stiles into the sill.

When all parts of the frame have been completed, but before final assembly takes place, the housings for the cord pulleys must be prepared near the top of the pulley stiles, and, so that the weights can be removed from the boxes when cords are being renewed, pocket pieces have to be cut near the lower ends of the stiles. Fig. 27 shows a portion of a pulley stile with the pocket piece removed. The pocket piece is the portion of the pulley stile which has been taken out so that the weights can be removed, and has a rebate at both ends, the one at the top having one of its edges bevelled. If a pocket piece is prepared in this way it will require only one screw to keep it in position.

Cutting the pockets. It should be cut in the following manner. When the pocket piece has been marked on the stile—the length of it will depend on the length of the weights—bore two $\frac{1}{4}$ in. or $\frac{3}{8}$ in. holes at the top and bottom marks in the groove for the parting bead. Using a pad-saw, cut down through the centre of the groove to the hole at the bottom. Lay the stile down on the bench face upwards, and with a fine tenon-saw make cuts half-way across and half-way through the thickness of the stile as in Fig. 28.

Turn the stile over on to its other face and make two more saw cuts half-way through its thickness as seen in the same drawings. Allow one end of the stile to rest on the bench face downwards, and, grasping the other end with the left hand, sharply rap a hammer near the top of the pocket piece and then near the lower end. The short lengths between each pair of saw cuts will split away leaving the pocket piece free. One screw, through the lower rebated portion, will be sufficient to hold it in place. The tongue on the side of the pocket piece should be removed.

Assembling the frame. Fig. 31 shows the first step in the assembly of the frame. The head, sill, and pulley stiles are all secured together as a frame using synthetic resin glue and nails. The inside and outside linings are fixed, these not usually being glued. Oval 2 in. brads are quite sufficient for this purpose. It is usual to place several glue blocks between the head and head linings to add strength to these parts (see Fig. 30). The parting slips should be fitted, odd pieces of plywood or hardboard usually being used. The back linings are fixed, being nailed only through the edges of the external vertical linings. The parting bead can also be fitted. No fixing should be necessary for these.

Fig. 30 shows a larger view of a vertical section through a boxed frame with sliding sashes. An anti-draught bead has been introduced in this, so that the lower sash can be raised slightly to obtain ventilation through the space between the meeting rails of the sashes.

Figs. 32, 33, and 34 show details of the sashes. The joints for the meeting rails are shown, and these can be of two kinds—either mortised-and-tenoned or dovetailed. The meeting rails have to be slightly wider than the other parts of the sashes to close the gap taken up by the parting bead. If this space were not closed when the sashes are in the closed position, a constant draught would be set up.

The meeting edges of the rails should be shaped as shown in Figs. 30 and 33. Meeting rails are usually kept a little narrower than the stiles and other rails to allow maximum light to enter. Consequently when the mortice-and-tenon joint is used at the meeting rail the joint usually takes up the full width of the rail. This requires a horn being left on as seen in Fig. 33, these being moulded to any required shape. If the dovetail joint is used no horns are required.

Grooves are worked into the edges of the stiles for about 12–15 in. from the top to receive the cords on to which the weights are fixed. Since these grooves considerably reduce the width of the stile as far

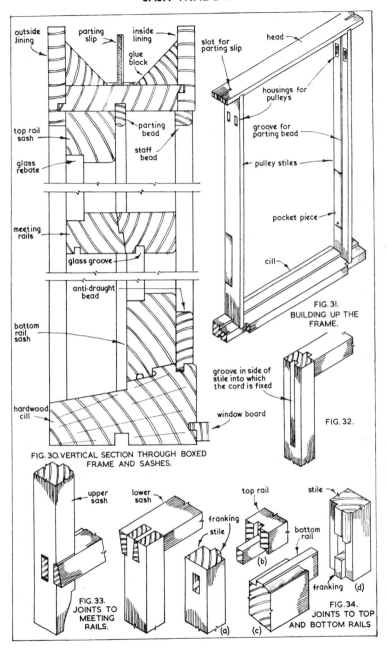

outside lining
parting slip
inside lining
glue block
slot for parting slip
head
top rail sash
parting bead
staff bead
glass rebate
housings for pulleys
groove for parting bead
pulley stiles
meeting rails
glass groove
pocket piece
anti-draught bead
cill

FIG. 31. BUILDING UP THE FRAME.

bottom rail sash

groove in side of stile into which the cord is fixed

FIG. 32.

hardwood cill
window board

FIG. 30. VERTICAL SECTION THROUGH BOXED FRAME AND SASHES.

upper sash
lower sash
franking
stile
top rail
stile
bottom rail

FIG. 33. JOINTS TO MEETING RAILS.

(a)
(b)
(c)
franking
(d)

FIG. 34. JOINTS TO TOP AND BOTTOM RAILS

as the mortice-and-tenon joints are concerned (see Fig. 32), it is necessary to dispense with the haunching. Instead, what is called a franking is used. In effect, the haunching is left on the stile and the housing for the haunching is made in the end of the rail, see Figs. 34 (*a*, *b*, and *d*). Where possible, all mouldings should be scribed as shown in Fig. 34.

Also see chapter 27.

DOOR FRAMES HELD SQUARE WITH TEMPORARY BRACES

CHAPTER 15   DOORS AND DOOR FRAMES

FOR THE PURPOSE of this chapter we deal only with the type of door found in and around the small domestic house, but it should be remembered that the principles involved in these can just as well be applied to the construction of more elaborate forms found in public buildings and the larger residence.

Ledged and braced door.   Fig. 1 shows the back elevation of a ledged and braced door, which is fairly cheap to produce and is suitable only for the doorways to sheds, coalhouses, and the like. It consists of narrow boards, usually matching, securely held together with three battens, called ledges.   When matching or tongued-and-grooved boards are used, the tongues and the grooves should first be painted with a good lead priming coat and two coats of paint of the finished colour.   Shrinkage of the boards usually takes place in the warmer months of the year, and the painting of the tongues and grooves not only prevents the penetration of moisture but also avoids the door becoming unsightly when part of the tongues can be seen.

The ledges are secured to the back surface of the door with nails which should be of the round wire type.   When driven through the face of the door the nails should be long enough for the points to pass right through the ledges and project at least $\frac{1}{4}$ in. beyond.   The heads should be well punched below the face of the door, and the door turned over so that the points can be turned over in the direction of the grain in the ledges.   They are then well punched below the surfaces.   The holes in the face of the door and the ledges can be filled with putty after the priming coat of paint has been applied.

If the ledged door were left like this the closing edge would probably drop slightly with its own weight, and so become out of square and unsightly.   It is possible that it would become difficult to open and close because the lower edge would scrape against the ground. To prevent this, braces are introduced.   Two of these can be seen in the drawing.   It is essential that the lower ends of the braces are always nearest the edge of the door which has the hinges fixed to it. Often these braces are cut to fit and butt against the lower and upper edges of the braces, but this is not a satisfactory way of fitting them.

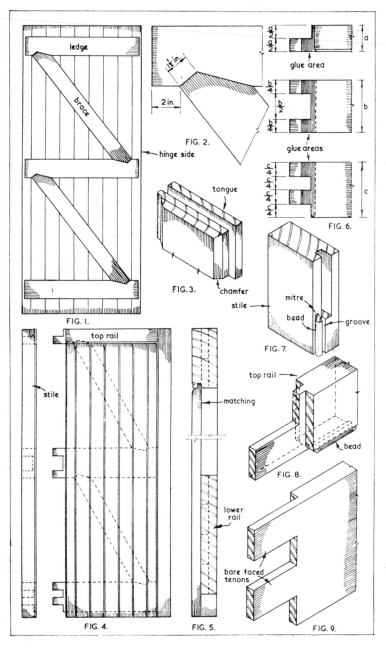

FIG. 1.

FIG. 2.

FIG. 3.

FIG. 4.

FIG. 5.

FIG. 6.

FIG. 7.

FIG. 8.

FIG. 9.

They are more capable of resisting the downward drag if they are housed into the edges of the ledges as shown in Fig. 2. The housings should be kept back at least 2 in. from the ends of ledges.

Framed, ledged and braced door. A more elaborate form of the ledged and braced door is that shown in Fig. 4. The panel is again formed with matched or tongued-and-grooved boards. The boards are framed, the thickness of the stiles and top rail being the required thickness of the door. The thickness of the middle and bottom rails and the braces are equal to that of the stiles less the thickness of the matching. Thus, if the finished thickness of the door is to be $1\frac{3}{4}$ in., and the finished thickness of the matching is $\frac{5}{8}$ in., then the thickness of the middle and bottom rails and the braces will have to be $1\frac{1}{8}$ in.

A glance at the elevation (Fig. 4) shows that the matched boards run in front of the middle and bottom rails and the braces. One of the stiles has been withdrawn so the mortice-and-tenon joints can be seen. The top rail (Fig. 8) which is equal to the full thickness of the door, has a single tenon and if, as is usually the case, a bead is worked on the inside edges of the stiles and top rail, the front shoulder has to be slightly longer than that at the back to allow for mitreing with the bead on the edge of the stile (see Fig. 7).

The middle and bottom rails of the door are thinner than the top rail because the matching has to pass in front of them. Consequently it is necessary to use bare-faced tenons where these two rails are jointed to the stiles. Fig. 9 shows one end of the bottom rail.

To obtain the strongest joints it is necessary to set out the mortice-and-tenon joints to any framed door so that the total width of the haunchings equals the total width of the tenons. Fig. 6 shows how this should be done. In the top rail the width a is the total of the widths of the tenon and the haunching. Half of this width should be occupied by the tenon and half by the haunching. The width b in the middle rail is the total width of the tenons and haunching. Again the total widths of the tenons should equal the width of the haunching. This rule applies also to the bottom rail. The width c is divided into four equal parts, two for the tenons and two for the haunchings.

It is important that this rule is followed for the middle rail for two reasons. The first is that the strongest joint possible is produced, and secondly, it may be necessary to fix a mortice lock and as it should be positioned along the centre line of the middle rail, the wide haunching between the two tenons will allow the mortice to be cut to receive the lock without cutting away any parts of the tenons.

Grooves are worked on the inside edges of the two stiles and top rail to receive the tongues along the top and down the outside edges of the matching. Fig. 3 shows the top end of one of the outside matched boards and illustrates how the tongues are worked on the appropriate edges of the matching. Chamfers, too, are often worked on the top and side edges of the matched boards and these are also shown in Fig. 3.

Panelled doors. There are many kinds of internal panelled doors, one of which is shown in Fig. 10. It consists of stiles, top and bottom rails, and intermediate rails, and $\frac{1}{4}$ in. plywood panels. The number of intermediate rails depends on the number of panels required in the door. To avoid the door looking top-heavy, the bottom panel may be made a little wider than the rest. The joints to the top and bottom rails are set out as before, first allowing for the panel grooves, and if the intermediate rails are not wider than 4 in. a single tenon as shown in the elevation can be used.

A moulding is usually worked on the inside edges of the framing, in which case allowances have to be made on the shoulders of the rails, or the framing can be made with the inside edges square and mouldings fixed round the edges of the panelling after the door has been assembled and the surfaces cleaned off.

Figs. 19 and 20 show these two finishes. Fig. 19 shows a small portion of the door-framing with a plywood panel in position in the grooves. Ovolo mouldings have been worked in the solid on the inside corners of the framing. These are termed "stuck" mouldings. Fig. 20 shows that the inside corners of the framing have been left square and moulding mitred and pinned around the panels at a later stage. These are termed "planted" mouldings.

Figs. 11, 12, and 13 show sections through three types of finishes, Figs. 11 and 12 having been already mentioned. Fig. 13 is termed a bolection moulding, and is similar to the planted moulding, but in this case the front edge of the moulding projects in front of the surface of the framework. Bolection mouldings are fixed by screwing through from the back of the panel. The screw holes in the panel should be slotted to allow for shrinkage and expansion of the panel. The screw holes are hidden by fixing planted mouldings on the other side of the door. These are fixed with pins, which should be driven through the moulding into the framework of the door and not into the panel as they would otherwise restrict the movement of the panel.

When stuck mouldings are used they should, where possible, be scribed at their intersections in preference to being mitred. Figs. 14 and 15 show how the mouldings of the top and intermediate rails are

PANELLED AND GLAZED DOORS

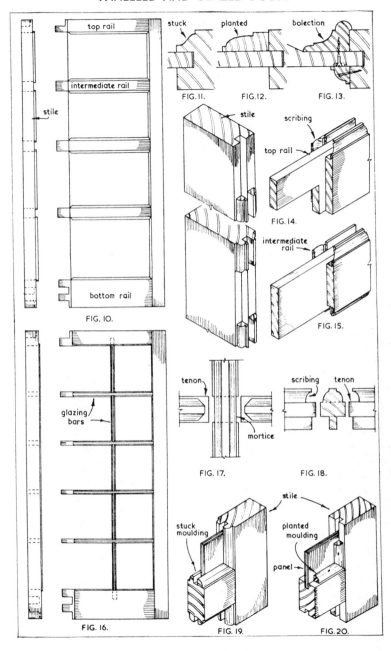

top rail

intermediate rail

stile

bottom rail

FIG. 10.

stuck

FIG. 11.

planted

FIG. 12.

bolection

FIG. 13.

stile

scribing

top rail

FIG. 14.

intermediate rail

FIG. 15.

glazing bars

FIG. 16.

tenon

mortice

FIG. 17.

scribing   tenon

FIG. 18.

stile

stuck moulding

planted moulding

panel

FIG. 19.

FIG. 20.

153

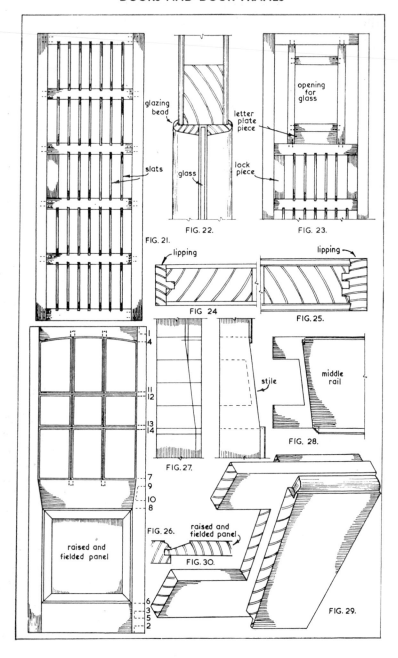

FIG. 21.

glazing bead

slats

FIG. 22.

glass

letter plate piece

lock piece

opening for glass

FIG. 23.

lipping

FIG 24

lipping

FIG. 25.

stile

middle rail

FIG. 28.

FIG. 27.

FIG. 26.

raised and fielded panel

raised and fielded panel

FIG. 30.

raised and fielded panel

FIG. 29.

scribed to the mouldings on the stiles. The top rail has been turned upside-down so that the scribed mouldings can be clearly seen.

Glazed door. Fig. 16 is the elevation of a fully glazed internal door which has ten glass panels, all of equal size. Notice that the tenons on the vertical bar are stubbed but those on the horizontal bars pass right through the stiles. These greatly strengthen the stiles which have no middle rails to support their centres. Figs. 17 and 18 illustrate the type of joint which is used at the intersection of the bars. The horizontal bars remain in one piece and have the mortises cut at their centres, and the vertical bar is cut into five lengths and the ends of these receive the tenons. The mouldings on the vertical pieces are scribed over those on the horizontal bars.

Flush doors. Flush doors have become popular in recent years and are being used for external purposes as well as for inside work. Mass-produced flush doors have for their framework timber of small section being, in some cases, as small as 1 in. by 1 in. Some manufacturers have their own special fillings for such doors as purpose-made wood shavings, corrugated cardboard, and many other materials in special forms.

However, it is sometimes necessary for the craftsman to make an individual or purpose-made flush door, and Fig. 21 shows one method which can be adopted. It can be faced with plywood sheets or hardboard, whichever is preferred, and the edges of the sheet material covered by gluing a strip of wood to the edges of the framing as in Fig. 24. These strips are called lippings. In ordinary work lippings are glued to the edges of the stiles only, but in good-quality work they are sometimes glued round the four edges of the door. So it is first necessary to ascertain how thick these lippings are to be and whether or not they are to cover the four edges of the door, because this will affect the setting out of the frame for the door. The thickness of the surface material must also be known before setting out the rod. For internal doors, $\frac{1}{8}$ in. plywood or hardboard is commonly used for covering the surfaces of the door, and as these doors are usually $1\frac{3}{8}$ in. thick, the frame thicknesses will be $1\frac{1}{8}$ in. thick (see Fig. 22).

External flush doors are often covered with $\frac{1}{4}$ in. resin-bonded plywood, and, as the finished thickness of external doors is commonly $1\frac{3}{4}$ in., the thickness of the frame should be $1\frac{1}{4}$ in. (see Fig. 25). It is also usual to have the lippings for external doors a little thicker than those for internal doors, say $\frac{1}{2}$ in. to $\frac{5}{8}$ in.

The elevation of a frame suitable for an internal or external flush door is seen in Fig. 21. It consists of two stiles and six rails all equal in width and thickness. Mortice-and-tenon joints are seldom used

155

FLUSH FRONT ENTRANCE DOOR WITH GLAZED PANEL
The frame also incorporates glazed areas above and at each side

for these frames.   A couple of $\frac{1}{4}$ in. or $\frac{3}{8}$ in. dowels are sufficient and sometimes corrugated dogs are used to keep the pieces together until the hardboard or plywood has been glued to the two faces. In addition to the framing already mentioned, $\frac{3}{8}$ in. slats are used between the rails, spaced at 3 to 4 in. apart to provide a glue surface. The slats fit into housings in the rails.

So that mortice or other locks can be fitted to the door an additional piece, similar to the framing, should be inserted half-way down the length of the door and adjacent to one of the stiles as shown in the drawing.   This will allow a mortice lock or a rim lock to be fitted to the door.   Many manufacturers insert a piece for a lock on each side of the door so that a mistake cannot be made, but if a door is made with only one of these pieces the side of the door which contains the lock piece should be clearly marked.

The lippings to the edges of the door should be tongued and fitted into a groove to provide a much more secure job, and to give a much larger gluing surface.   The external door lippings should be fixed in a more elaborate manner, as seen in Fig. 25.   There are two tongues, and the edges of the plywood fit into the bevelled edges of the lippings.   If the lippings cover all four edges of the door they are mitred together at the corners.

Sometimes external flush doors have glass panels and also a means of fixing a letter-plate.   One method is shown in Fig. 23.   The top portion of the door has additional framing and this is secured to give the required glass sizes.   At the bottom of the additional framing another piece of timber has been inserted to provide a means of fixing the letter-plate.   Fig. 22 is a larger drawing through one edge of the glazed portion of the door, and shows the glazing beads on each side of the glass pane for securing it in position.   These are fixed with panel pins.

Glazed entrance door.   More elaborate forms of external glazed doors are often required as shown in Fig. 26.   The stiles are reduced in width at the glazed portion of the door and the lower edge of the top rail is segmental in shape.   Stuck mouldings are worked on the front edges of the framing around the panelled portion of the door and the panel is raised and fielded, as seen in Fig. 30.

Fig. 27 illustrates how the stile is marked out for the joint to receive the tenons on the end of the middle rails.   Fig. 28 shows that portion of the stile and the end of the middle rail prepared ready for assembly.   Fig. 29 is a pictorial view of one end of the middle rail.

Assembling doors.   When doors are ready for gluing up they should be placed preferably on a bench specially made for this purpose.   Failing this, two fairly large pieces of timber, say, 4 in. by

4 in. should be laid across a joiner's bench and, if necessary, packed up so that their top surfaces are in line with each other. The door is laid on top of the two pieces and the stiles tapped away from the shoulders so that the ends of the tenons are in the mortises about 1 in. Glue is then applied on both sides of the portions of the tenons shown in Fig. 6, being the half nearest to the shoulder. If this is observed the glue will not restrict the stiles from shrinking or expanding.

When the glue has been applied, the stiles should be tapped towards the shoulders with a hammer used over an odd piece of wood to avoid bruising the edges, and the cramps applied. Before driving the wedges home the squaring rod should be placed across the diagonals of the door to check for squareness. To make sure that all the joints pull up and the rails are positioned correctly it is wise to drive the wedges home in the order shown to the right of Fig. 26. There are fourteen wedges on each side of the half-glazed door and it will be seen that the outside wedges of the top and bottom rails should be driven first.

Synthetic resin glues should be used for doors, especially external doors. These are much better than the outdated Scotch glue which is affected by dampness.

Internal door linings. Frames for internal doors are normally made from fairly thin pieces of timber and are called linings. Fig. 31 is the elevation of a lining made from timber 1 in. to $1\frac{1}{2}$ in. thick. A section through a lining is shown in Fig. 32, and all it requires is a rebate worked on one edge, the width equal to the thickness of the door. The two vertical linings are tongued and grooved to the head or horizontal lining, and this joint is seen in the exploded details shown in Figs. 33 and 34. The joints are glued and nailed and a distance-piece is fixed near the feet of the vertical linings as shown. This should be fixed carefully so that it is an equal distance from the end of each of the feet as it is to be used in the squaring of the frame.

A squaring rod, which is a piece of timber about 7 ft. in length and 1 in. square in section is employed in checking the frame for squareness and this is done by measuring the diagonals of the frame shown with broken lines in Fig. 31. This is the reason why the distance-piece has to be fixed in its position very carefully. If the two diagonals are equal in length the framing is square. To hold it square until it is required for fixing, braces can be nailed across the corners of the linings as shown in the drawing.

The fixing of the linings is dealt with in another chapter, but Fig. 35 shows the linings in position in the door-opening with architraves covering the joint between the linings and plaster on the faces

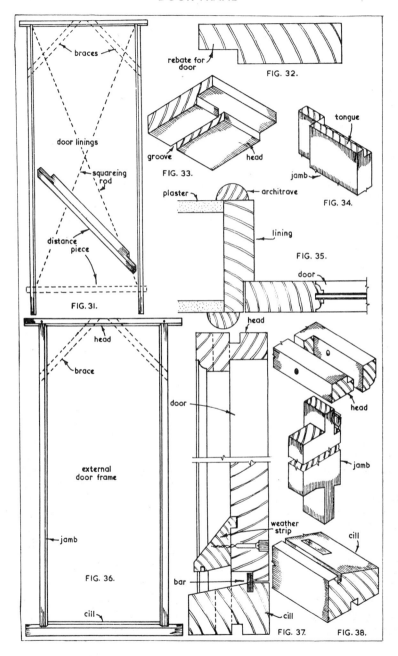

braces

rebate for door

FIG. 32.

door linings

groove

head

FIG. 33.

tongue

jamb

FIG. 34.

squareing rod

plaster

architrave

lining

FIG. 35.

distance piece

door

FIG. 31.

head

head

brace

door

head

external door frame

jamb

weather strip

cill

jamb

FIG. 36.

bar

cill

FIG. 37.

cill

FIG. 38.

159

DOOR FRAMES IN POSITION. NOTE TEMPORARY BRACES

T.D.A. HALF TRUSS IN POSITION

of the partition.   It should be noted, therefore, that the width of the linings must be equal to the thickness of the partition and the two thicknesses of the plaster on the surfaces of the partition.

External frame.   Fig. 36 shows the elevation of an external door frame and is much more robust than the linings.   The timber normally used for the head and posts, or jambs, is between 3 in. by 2 in. and 4 in. by 3 in.   The sill, which should be made from timber such as English oak or some other durable hardwood, should be 2 to 3 in. thick and a little wider than the posts (see Fig. 37).

External doors open into the building and every effort should be made to make them draught and waterproof.   The sill must be weathered, the bevel being about $\frac{1}{2}$ in. in depth.   A mild steel bar should also be let into the top surface of the sill, and a rebate worked on the bottom edge of the door should fit up tight against the bar when the door is closed.   In addition to this an oak weather strip should be fixed, preferably tongued, grooved, and screwed near the bottom of the door.   The groove and tongue should be well painted before the strip is fixed.

Fig. 38 shows, in isometric, the joints used for the frame.   These should be secured with a synthetic resin glue and hardwood dowels.

Also see chapter 25.

CHAPTER 16   SPLAYED WORK

THE DEVELOPMENT OF inclined surfaces occurs so often in carpentry and joinery that a sound knowledge of the geometry of inclined or splayed work is essential to the craftsman following these trades.   Roofing work, linings round door- and window-openings, and form-work for mushroom-headed concrete columns are only a few examples where these problems are likely to occur.

Inclined plane.   Fig. 1 is the plan and elevation of an inclined surface a–b–c–d.   Its inclination can be seen in the elevation, and the plan of the surface, a–b–c–d, can be seen below the elevation.   The problem is to develop the shape of the surface.   With the compass point in a, in the elevation, and radius a–b, describe the arc to give point b″ on the X–Y line.   Drop a vertical line from b″ to meet horizontal lines brought over from points c and b in the plan to give c′ and b′.   a–b′–c′–d is the true shape of the surface a–b–c–d.

Although the surface can be seen in the plan, this is not its true shape because it is inclined from the edge a–d upwards at an angle, in this case, 30 degrees.   To obtain its true shape we must imagine that the surface is being laid down flat on the X–Y line.   If this is done the point b in the elevation follows the arc drawn with the compasses and would come to rest at b″.   Viewing this movement from above, b and c in the plan would move towards the right and come to rest immediately below point b″ to give the points b′ and c′.   Notice, too, that the points b and c move outwards at right angles to the hinge line a–d.

Splayed work.   When developing any inclined plane with this method, all points which move sideways or downwards or, in fact, in any direction, must always move at right angles to the hinge line. This is evident in the second series of drawings (Fig. 2).   These show the plan and elevation of two inclined fascia boards to a shop front and intersect at an angle of 90 degrees.   The two boards a–b–c–d and a–b–e–f are mitred together at a–b.   The problems here are to develop the shapes of the two boards and also the bevel to apply to the ends so that they mitre together correctly.   As the elevation shows, the boards are inclined at an angle of 60 degrees.

To develop the boards, place the compass point in b in the elevation, and with radius b–a describe the arc to give a′ on the

DEVELOPING SURFACES

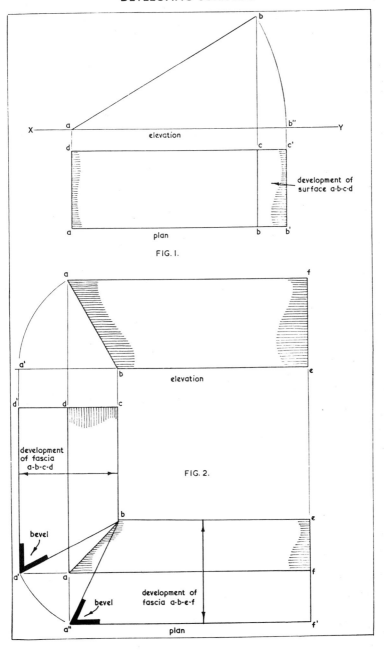

elevation

b''

X — a — Y

d c c'

development of
surface a-b-c-d

a b b'

plan

FIG. I.

a f

elevation

a'

b e

d' d c

development
of fascia
a-b-c-d

FIG. 2.

bevel

b e

f

a' a

bevel

development of
fascia a-b-e-f

a'' f'

plan

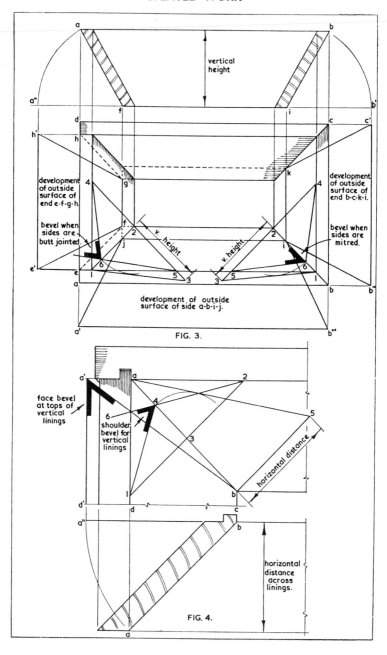

vertical height

development of outside surface of end e-f-g-h.

bevel when sides are butt jointed.

development of outside surface of end b-c-k-i.

bevel when sides are mitred.

v. height

v. height

development of outside surface of side a-b-i-j.

FIG. 3.

face bevel at tops of vertical linings

shoulder bevel for vertical linings

horizontal distance

horizontal distance across linings.

FIG. 4.

horizontal line brought out from b. Drop a vertical line downwards from a′ to give points a′ and d′ in the plan. a′–b–c–d′ is the development of one of the boards. In this case b–c is the hinge line, and so points a and d must travel outwards at right angles to this line.

The width of the other board, a–b–e–f, is exactly the same width as the first, so it is quite a simple matter to draw the true shape of this surface. The bevel required (both boards require the same one) is equal to the angle of the mitred ends of developed boards.

Fig. 3 is the plan and elevation of four sides of a box, all inclined at an angle of 60 degrees. The joints at the corners on the left-hand side of the box are butt joints, and those on the right-hand side are mitres. The problems are to develop the shapes of the sides and the bevels to make the joints at the corner. The end of the box to the left of the plan has its outside surface marked e–f–g–h, and to develop this surface the point of the compasses is placed in point f in the elevation and with radius f–a describe the arc to give point a″ on the horizontal line brought out from the base of the elevation. From a″ drop a vertical line downwards to intersect with two horizontal lines brought out from points e and h in the plan to give points e′ and h′. e′–f–g–h′ is the developed outside surface of the side.

At the opposite end of the box the corners are mitred. The outside surface in this case is b–c–k–i, seen in the plan. To develop this surface, place the compass point in i in the elevation and with radius i–b describe the arc b–b′. Drop a vertical line from b′ to give points c′ and b″ on horizontal lines brought out from points c and b in the plan. The developed surface is b″–c′–k–i. The development of the outside surface of one of the longer sides of the box is seen on the lower end of the plan.

To develop the outside surface of side a–b–i–j drop vertical lines from points a and b in the plan and with compass point in i and radius i–b″ describe an arc to give point b‴ on the vertical line drawn from b. Draw a horizontal line from b‴ to give point a′ on the vertical line dropped from point a. a′–b‴–i–j is the developed outside surface of the longer side of the box. The outside surface of the other long side is the same as the one developed.

Whether the corners are mitred or butt jointed a bevel has to be applied to the ends of the pieces so that each piece will fit against the surface of the adjacent piece correctly. If the corner is mitred, the ends of both pieces have to be bevelled, but if butt jointed a bevel is applied to one of the two pieces only.

Mitred corner. Let us consider the mitred corners first. To obtain the mitre bevel the dihedral angle of one of the corners must be developed. The dihedral angle is the angle made by two adjacent sides, and involves developing the shape of a triangle which will just fit on the two inside surfaces of the corner with the surface of the triangle at right angles to the corner line.

To develop the dihedral angle for the corner at b:

1. Obtain the true length of the corner 1–2 by drawing line 2–3 at right angles to 1–2. Make the length of 2–3 equal the vertical height of the box. Line 1–3 equals the true length of 1–2.

2. Extend 2–3 across to give point 4 on the inside edge of the adjacent side of the box.

3. With compass point in point 2 and the compasses open to just touch 1–3, describe an arc to give point 6 on line 1–2.

4. Join 6 to 4 and 6 to 5 with straight lines. Angle 4–6–5 is the dihedral angle of the corner. The mitre bevel is exactly half of the dihedral angle and is 2–6–5.

Butt joint. To obtain the bevel to apply to one of ends when the joint at each corner is a butt joint, the dihedral angle is developed as before and the external angle set up by the dihedral is the bevel required (see corner a).

Fig. 4 is another example where the dihedral angle has to be developed to obtain the bevel for a butt joint to a piece of work with splayed sides. In this case the drawings illustrate a corner of some splayed linings round a door or window-opening. The joint shown is a tongued-and-grooved joint but the shoulder to the tongued portion is the same as a butt joint and so the same geometry applies.

The inside surface of the vertical lining is developed in a similar way as in the previous drawings. With compass point in b in the plan and radius b–a, describe an arc to give point a″ on a horizontal line brought over from b. From a″ draw a vertical line upwards to intersect with horizontal lines brought over from a and d in the elevation to give points a′ and d′. a′–b–c–d′ is the developed surface and angle b–a′–d′ the bevel to apply across the wide surfaces of the two vertical linings.

The dihedral angle is developed by first finding the true length of the corner a–b. Make a–b–5 a right angle and b–5 equal to the horizontal distance across the linings. a–5 is the true length of a–b. Draw 1–2 at any point so that it is at right angles to a–b. Then with compass point in 3 and the compasses open to just touch a–5, describe an arc to give 4 on a–b. Join 2 and 4 and then 1 and 4 with straight lines. Angle 1–4–2 is the dihedral angle. Angle 1–4–6 is the bevel required.

COMPOUND CUT AT TOP END OF JACK RAFTER

SOME OF THE SIMPLE AND COMPOUND CUTS IN ROOF WORK

CHAPTER 17 STAIRCASE WORK

BEFORE EMBARKING ON the construction of a staircase it is advisable to visit the site where the work is to be fixed to check the various dimensions required.

Checking sizes. The most important of these dimensions is the vertical height between the floors. Failure to check this measurement may result in the height of the top step being different from the rest and thus becoming a trap or, alternatively, the stairs being fixed at the wrong angle. The distance between the walls of the stair well and the positions of doors and windows, the height of landings, etc. too must all be checked and noted for future reference.

The vertical heights between floors and also floor and landing are usually marked on a piece of timber about 1 in. square in section so that no possible mistake can be made. This piece of timber is called a storey rod, and one of these can be seen in Fig. 1. It is the storey rod to the short flight of stairs in Fig. 2. On the rod has been marked the thickness of the floor-boards of both floors and also the position of the trimmer-joist of the landing.

When the number of steps to the flight has been decided upon, the vertical height of the storey rod will be divided up into that number of equal spaces. This will ensure that each step has the same rise as all the others. The 'rise' of a step is the vertical height between the top surface of its tread down to the top surface of the tread immediately below it. The 'going' of a step is the horizontal distance between the front surface of its riser to the front surface of the riser of the next step (see Fig. 3).

Tread and riser ratio. Many people consider that there should be a ratio between the tread and riser of each step, and they base their calculations on the length of a pace of the average individual. This is said to be 23 to 24 in. A formula used for working out the ratio is: twice the rise in inches plus the width of tread in inches equals 23 to 24 in., the reasoning being that it takes twice as much effort to go upwards as it does to go forwards.

For example, supposing the height of each step has been calculated as being 7 in. (this is obtained from the storey rod) the ideal width of tread would be found thus:

$$23 \text{ in. minus } (7 \text{ in.} \times 2) = 9 \text{ in.}$$

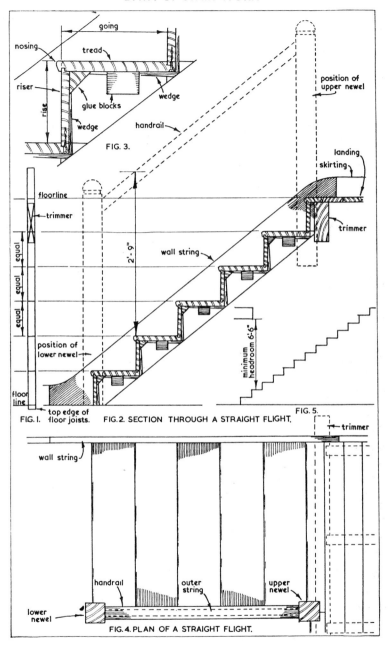

going

nosing

tread

riser

rise

glue blocks

wedge

wedge

handrail

FIG. 3.

position of
upper newel

landing

skirting

floorline

trimmer

trimmer

2'-9"

wall string

equal

equal

equal

position of
lower newel

minimum
headroom 6'-6"

floor
line

top edge of
floor joists.

FIG. I.

FIG.2. SECTION THROUGH A STRAIGHT FLIGHT.

FIG. 5.

wall string

trimmer

handrail

outer
string

upper
newel

lower
newel

FIG. 4. PLAN OF A STRAIGHT FLIGHT.

PREFABRICATED STAIRCASE CONSTRUCTION

Most staircases are made in the workshop and assembled in sections which can be dismantled for easy transport to the site. They are then erected and fitted in the building.

Photograph by courtesy of Austins of East Ham Ltd.

The ideal going for a step with a rise of 7 in. is, therefore, 9 in. Obviously, this method for finding the going and rise of a staircase cannot always be followed, but it does give a guide as to what the ratio should be.

Straight flight. Now let us consider the short, straight flight of stairs shown in Figs. 2 and 4. The broken lines to the right of the floor, Fig. 4, are the positions of the 8 in. by 3 in. trimmer-joist with 8 in. by 2 in. trimmed joists housed into the trimmer, the joint being similar to those shown in Fig. 23, Chapter 3, or Fig. 5, Chapter 12. These joists will form the landing at the top of the flight.

The plan also shows that the flight has a wall string and an outer string. The steps are housed to a depth of about $\frac{3}{8}$ in. into the strings. The outer string is morticed and tenoned to a 4 in. by 4 in. newel post at each end. The outline of the wall string is clearly seen in Fig. 2 and this shows that the string is shaped at the lower end so that it will intersect with the skirting board on the lower floor. The drawing also shows that the wall string is cut to fit over the top edge of the trimmer at the landing, and also shaped to intersect with the skirting on the landing.

The positions of the newels should also be noted carefully. The front edge of the first riser is on the centre line of the lower newel, and the front edge of the top riser is on the centre line of the upper newel. The newels must be long enough to enable a handrail to be morticed and tenoned to them so that its top edge is approximately 2 ft. 9 in. from the nosing of each step, measured vertically. The height of a hand-rail on a landing should be approximately 3 ft. 1 in. Baluster rails are square or moulded pieces of timber morticed and tenoned to the top edge of the outer string and the lower surface of the handrail.

Steps. Details of the steps are seen in Fig. 3. These are made up as steps on the bench and stacked until required. The risers, in this detail, are tongued on both edges, and the treads grooved, top and bottom as seen in the drawing. The steps are housed into the strings to a depth of up to $\frac{3}{8}$ in., and when the flight is assembled they are glued and wedged into the housings. Additional strength is obtained by gluing and nailing triangular blocks in the angles. At least three of these blocks should be fixed across the width of the step. Screws, too, are also used for securing the treads and risers of adjacent steps when the flight is assembled (see Fig. 3). It should also be observed that the width of the tread is greater than the going.

String and newel joint. Fig. 10 shows the type of mortice and tenon joint used for joining the outer string to the newels. This joint is used at the lower end of the flight and shows that the whole of

STAIRCASE WORK

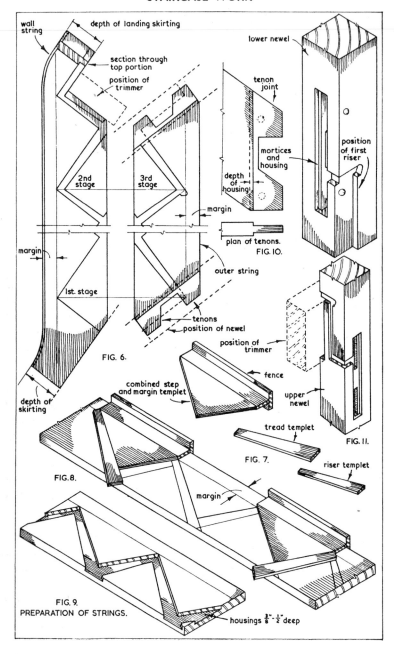

wall string

depth of landing skirting

section through top portion

position of trimmer

2nd stage

3rd stage

1st. stage

margin

margin

depth of skirting

FIG. 6.

lower newel

tenon joint

mortices and housing

depth of housing

position of first riser

plan of tenons.
FIG. 10.

outer string

tenons
position of newel

position of trimmer

fence

combined step and margin templet

tread templet

upper newel

FIG. 11.

FIG. 7.

FIG. 8.

riser templet

margin

FIG. 9.
PREPARATION OF STRINGS.

housings $\frac{3}{8}" - \frac{1}{2}"$ deep

the string is housed into the newel to a depth of about $\frac{1}{4}$ in. to avoid any unsightly gaps appearing due to shrinkage. The thickness of the tenon can well be more than one third the thickness of the string, a suggested thickness being $\frac{3}{4}$ in. The joint is glued and a $\frac{3}{8}$ in. dowel draw-bored through each tenon.

Fig. 10 also shows the preparation required by the newel to receive the bottom step. The riser can be secured by two screws from the back into the newel. Fig. 11 shows the cutting of the upper newel to receive the tenons of the string, and also the top steps of the flight. It shows, too, that the newel has been notched over the front edge of the trimmer. This helps the stairs to carry the weight to which it is subjected, and also adds enormously to their stability.

At least three templets are required for marking out the steps on the surfaces of the strings, and these are seen in Fig. 7. Often the margin is marked with a separate templet, so requiring four templets instead of three. First of all the margin, the distance between the point of the step to the top edge of the string, has to be ascertained. This can be any measurement, depending on the person designing the work. The combined step and margin templet with its fence secured to the top edge will enable the person marking out to make an outline of the steps similar to that shown at the lower end of the wall string and marked " first stage " (Fig. 6). The tread and riser templets must be shaped to allow for the wedge spaces in the housings and, as it is necessary to have the wedges fitting their spaces accurately, a specimen wedge should be obtained and the templets shaped to suit the wedges to be used.

The second stage (Fig. 6) shows that the housings have been marked with the aid of the tread and riser templets, and Fig. 8 will show this operation in progress. Fig. 9 shows the housing almost completed. To complete the housings the nosings on the steps must be marked and removed to the same depth as the other parts of the housings. It is usual to do these as a separate operation.

The steps which are already assembled and stacked ready for use should be numbered, number 1 being the bottom step, number 2 the next step upwards, and so on. When marking the shapes of the nosings on the strings the actual step should be used to mark the nosing on the housing into which it will be finally fixed. This is because there may be slight variations in the shapes of the nosings, especially if they have been shaped on the bench by hand. Fig. 6 shows the shapes of the wall and outer strings.

Dog-legged stair. Fig. 12 is the plan of a dog-legged stair which comprises of two flights, the outer string of the top flight being immediately above the outer string of the lower flight. The lower flight consists of six steps leading to a half-space landing which has been formed by building in a 3-in.-thick trimmer-joist and 2-in.-thick trimmed joists at approximately 18 in. centres.

The second, or upper flight has eight steps leading from the half-space landing up to the first-floor level. The two outer strings are morticed and tenoned to the landing newel in the manner shown in Fig. 13. As with the joint in Fig. 10 the strings and newel are glued together and draw-bored with ⅜ in. dowels. The position of the lower newel is slightly different from that in Fig. 2 because the first step is shaped at the outer end. These are dealt with later in the chapter. The front edge of riser number 2 is in line with the centre of the newel post in this and similar staircases.

The heights of the handrails are also shown in Fig. 14. That on the flight should have a vertical height of 2 ft. 9 in. above the nosing of each step, and that on the landing should be approximately 3 ft. 1 in. So that the handrail and balustrading can be kept central with the newel posts, it may be necessary for the edge of the landing to overhang the 3 in. joist anything up to 2 in. (see Fig. 14). The board forming the edge of the landing can be supported by tapered brackets, nailed or screwed to the edge of the joist and faced with hardboard or plywood.

The joints for securing the handrails to the newel posts are seen in Figs. 15 and 16. As with the strings, the whole of the handrail should be housed into the newel to a depth of about ¼ in. so that if shrinkage does occur no unsightly open joints will be seen.

Shaped steps. A method of constructing shaped steps is seen in Figs. 19, 20, and 21. Two common steps with shaped ends are the bull-nosed step, seen in Fig. 19 and at the lower end of the dog-legged staircase (Fig. 12), and the half-round step (Fig. 20) and also seen as the first step in Fig. 17.

Blocks are used in this method, and these are usually built up from timber of any thickness to the required depth. The block should have the grain of adjacent pieces running at right angles to one another and these are glued and screwed together, the finished thickness of the blocks being the depth of the step less the thickness of the tread. The shapes of the blocks required for a bull-nosed step and a half-round step are seen in Fig. 21. The riser of the step is first reduced to a thin veneer, usually about ⅛ in. where it is passed round the block (see Fig. 21). The end of the veneered portion of the riser nearest the newel post has a bevelled shoulder, and this portion of the riser fits into a recess cut into the shaped block.

Before assembling the step, the veneered portion of the riser must be steamed or soaked in hot water to soften its fibres. This will

DOG-LEGGED STAIRCASE

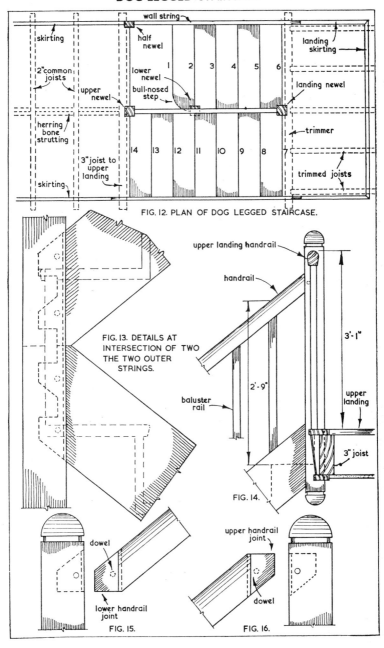

FIG. 12. PLAN OF DOG LEGGED STAIRCASE.

FIG. 13. DETAILS AT INTERSECTION OF TWO THE TWO OUTER STRINGS.

FIG. 14.

FIG. 15.

FIG. 16.

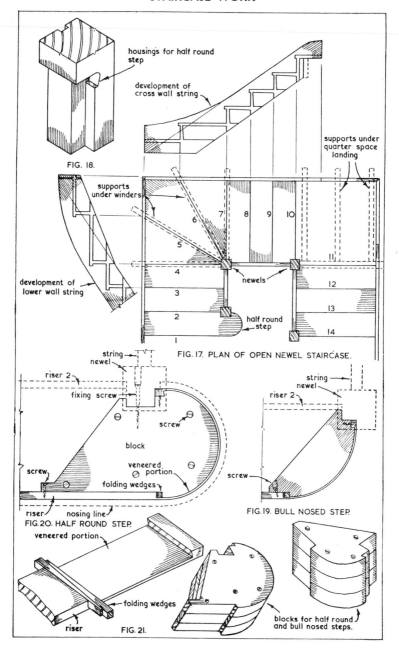

housings for half round step

development of cross wall string

supports under quarter space landing

FIG. 18.

supports under winders

development of lower wall string

newels

half round step

FIG. 17. PLAN OF OPEN NEWEL STAIRCASE.

string
newel

riser 2

fixing screw

screw

block

veneered portion

screw

folding wedges

riser   nosing line

FIG.20. HALF ROUND STEP.

string
newel

riser 2

screw

FIG.19. BULL NOSED STEP.

veneered portion

folding wedges

riser   FIG. 21.

blocks for half round and bull nosed steps.

allow it to be bent round the curved portion of the block. The end of the riser with the bevelled shoulder is secured in its recess with gluing and screws. The curved portion of the block and the inside surface of the veneer are glued, and the riser bent round the curve of the block.

A pair of folding wedges is used to stretch the veneer round the block, and these should also have glue spread over their surfaces before being inserted into the space provided for them. Care must be taken when cutting these wedges so that they fit perfectly in their place. Their length should be at least 3 in. more than the depth of the block. Their ends are cut off flush with the surfaces of the block after assembly. Two securing screws are inserted through the block into the riser at the end nearest to the folding wedges.

Fig. 18 shows the housing necessary in the lower end of a newel post to accommodate the block, and tread to a half-round step.

Open newel staircase. Fig. 17 shows the plan of an open newel staircase, which, of course, is different from the dog-legged stairs because a well is formed between the various flights. This illustration is only one example of an open newel staircase, and the 180-degree turn consists of a quarter-space landing between risers 10 and 11, a short straight flight of four steps with numbers 7 to 10 risers and a quarter-turn of winders. The stairs at this point, could, if so wished, have a half-space landing, or two quarter-turns of winders with a small landing between, and so on.

To the left and at the top of the plan are shown the shapes of the lower and cross wall strings. The extra width required in these strings is obtained by jointing another short length of board on the appropriate edge of the string and shaping in the way shown in the drawings. The quarter-space landing is supported by a 3-in.-thick trimmer-joist built into the wall at one end, and housed into the newel post at the other. The 2 in. trimmed joists are also built into the wall and supported at their other ends by notching into the trimmer. The newels would extend down to floor level.

Support is given to the winders by placing a 3 in. by 2 in. or larger piece of timber under each riser as seen in the drawing. They also have one end built into the wall, and the other end of each is housed into the newel post. A better idea of how this is done is seen in Fig. 22. The supports to the risers of the winders are seen housed into the newel post.

Fig. 23 shows how the marking out of the newel post which receives the ends of the winders is done. First a full-size plan of the newel and the positions of the front edges of the riser boards is drawn. This is seen at the bottom of the drawing. The newel can then be

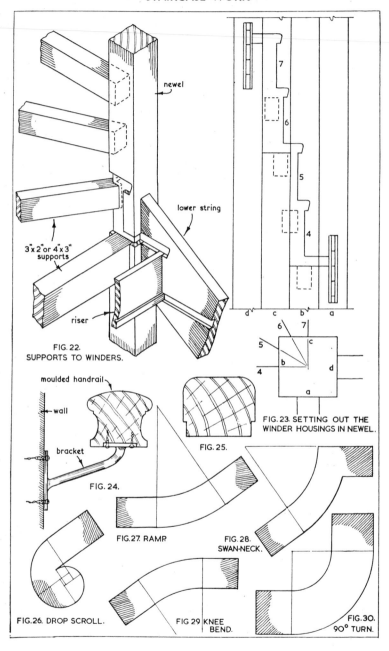

newel

lower string

3"x2"or 4"x3" supports

riser

FIG. 22.
SUPPORTS TO WINDERS.

FIG. 23. SETTING OUT THE WINDER HOUSINGS IN NEWEL.

moulded handrail

wall

bracket

FIG. 24.

FIG. 25.

FIG.27. RAMP.

FIG. 28. SWAN-NECK.

FIG.26. DROP SCROLL.

FIG 29 KNEE BEND.

FIG.30. 90° TURN.

set out on its four faces starting at the position of the tread to step number 3, the height of which can be easily calculated. Remember that the rise of each step is the same as all the others, and be sure that the position of the front face of each riser is placed in the correct position on the correct face of the newel post. The broken lines show the positions of the riser supports.

Fig. 24 shows a method of fixing a handrail to a wall by means of purpose-made brackets. Fig. 25 is a section through a simple handrail and Figs. 26–30 show various deviations from straight handrails. For instance the drop scroll (Fig. 26) could well be the lower end of the wall rail shown in Fig. 24. It is quite a satisfactory way of terminating a straight handrail of this sort. Figs. 27, 28, and 29 are often used at the lower or upper end of a rail just prior to its termination at a newel post. The 90-degree turn (Fig. 30) is used at a change of direction where no newel post exists. This, of course, is a plan, whereas Figs. 26, 27, 28, and 29 are elevations.

Methods for setting out scrolls and handrails and geometrical stairs will be fully covered in an advanced volume.

Also see chapter 31.

CHAPTER 18   SHORING

WHEN BUILDINGS ARE being repaired, alterations to be carried out, or if a building is in danger of collapsing due to damage or removal of adjacent property, temporary supports must be given to the building so that the work to be done can be carried out in safety to workmen and public alike, and also so that further damage to the building will not result.

If the shoring of buildings is to be carried out in an efficient manner, the person in charge of the work must have a fairly good knowledge of how buildings are constructed, and he must also know the best type of supports necessary for any particular job and how these supports should be positioned and assembled.

Walls which tend to bulge or lean outwards may be supported in two ways; by raking shores or flying shores.

Raking shores. Raking shores are supports which have their lower ends resting on the ground and their upper ends resting against the wall which is to be supported. An illustration of this type of support is shown in Fig. 4. The support, or raker, as it is often called, has its lower end resting against a base support, and its upper end bedding up against a wall piece which spreads the thrust of the raker over a large area of the wall.

Now, what should the position of the raker be so that it will give adequate support to the wall, which, say, is tending to bulge outwards? Let us first consider the lower end of the support. The more upright the raker is, the less chance it has of stopping the wall from moving outwards. The farther the foot of the raker is away from the wall the more efficient it becomes. But of course, in most jobs like this, the space available at the ground level of a building is limited. In roads, for instance, the width of the pavement is all the space one is able to use, unless conditions prove otherwise.

The position of the top end of the raker is also important. Fig. 1 shows how the position depends on the construction of the floors of the building, as it is usual practice to have the tops of the rakers around the positions of the floor levels. Where the floor-joists rest on the wall to be supported the centre line of the raker should be in line with the centre of the plate or seating.

RAKING SHORES

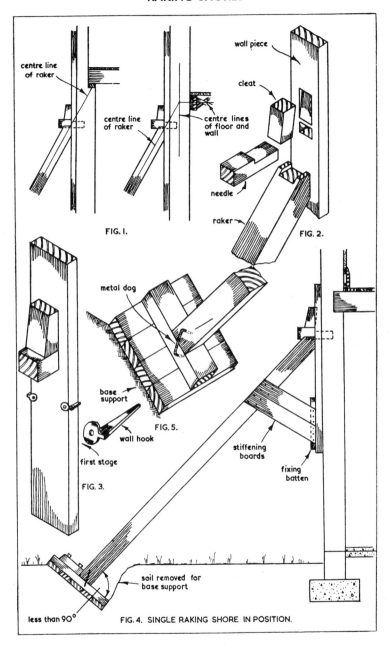

centre line of raker

centre line of raker

centre lines of floor and wall

FIG. I.

wall piece

cleat

needle

raker

FIG. 2.

metal dog

base support

first stage

FIG. 5.

wall hook

FIG. 3.

stiffening boards

fixing batten

soil removed for base support

less than 90°

FIG. 4. SINGLE RAKING SHORE IN POSITION.

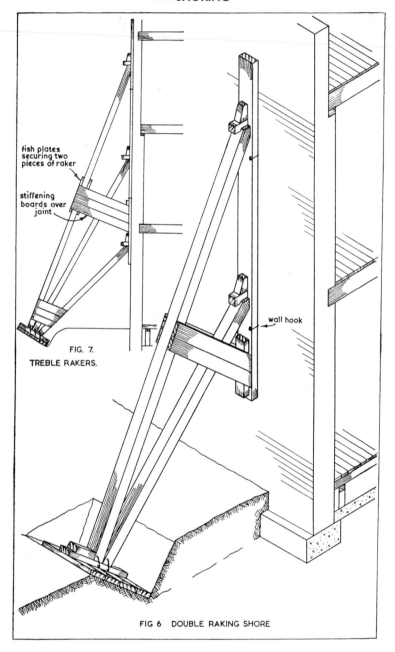

fish plates
securing two
pieces of raker

stiffening
boards over
joint

FIG. 7.
TREBLE RAKERS.

wall hook

FIG 6 DOUBLE RAKING SHORE

The second of the drawings in Fig. 1 shows that the floor-joists run in the same direction as the wall. In a case such as this, the centre line of the raker, the centre line of the floor, and the centre line of the wall should all meet at one point. If the raker can be placed opposite to a row of strutting so much the better.

Raker top joint. Fig. 2 shows an exploded view of the joints used at the head of the raker. A mortice, 4 in. by 3 in. in size is cut in the wall piece to receive the end of the needle, and a recess is also cut above the mortise to house the back edge of the cleat. The cleat and the needle overcome the upwards thrust of the raking shore, and the wall piece helps to spread that thrust over a large area of the wall.

Method of erection. To erect a system of raking shores, the positions of the needles must first be ascertained, and half-bricks (4 in. by 3 in.) removed from the wall at these points. Mortices are cut through the wall piece to coincide with these positions, and after the recessing has been completed above the mortices, the needles and cleats are assembled in the wall piece and fixed by nailing.

The assembled pieces are offered up in position on the wall, the ends of the needles entering the holes in the wall made by the removal of the half-bricks. The ends of the needles should enter the wall to a depth of about 4 in. The wall piece can then be fixed in position with the use of wall hooks, Fig. 3.

The position of the foot of the raker being prepared, the base supports should be placed so that the angle between the support and the inside edge of the raker is a little less than 90 degrees, so that the raker can be tightened up into its position by gently levering it with a crowbar towards the wall. When it has been tightened it should be secured in that position by using metal dogs and a cleat behind the foot of the raker. Fig. 5 is an isometric detail of the base support. The cleat has not been included.

Stiffening boards should be employed to give additional strength to the raker, and to assist in giving stability to the whole of the structure.

Double raking shore. Fig. 6 illustrates a double-raking shore, and shows the wall hooks which have been used for fixing the wall piece, the excavated ground for correct positioning of the base support, and the stiffening boards giving stability to both rakers.

Treble rakers. Fig. 7 shows three rakers used for the support of a four-storied building. It may not always be possible to obtain timber of required lengths. This can be overcome by joining two pieces together with fish-plates. An experienced man in charge of the shoring of a building will know what size timbers to use for such

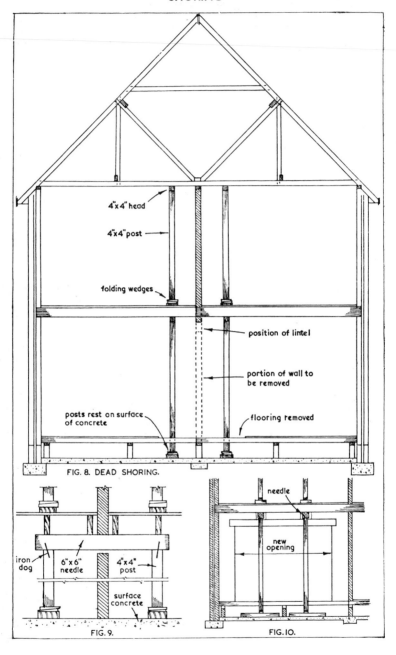

4"x 4" head

4"x 4" post

folding wedges

position of lintel

portion of wall to be removed

posts rest on surface of concrete

flooring removed

FIG. 8. DEAD SHORING.

iron dog

6"x 6" needle

4"x 4" post

surface concrete

FIG. 9.

needle

new opening

FIG. 10.

jobs, and the sizes will, of course, differ from job to job, but for straightforward shoring jobs for medium-size buildings the following timbers can be used.

Wall pieces from 6 in. × 2 in. to 9 in. × 3 in.

Rakers from 9 in. × 3 in., 5 in. × 5 in. to 8 in. × 8 in.

Base supports 3 in.-thick timbers.

Cleats 4 in. × 4 in.

Needles 4 in. × 4 in.

Dead shoring. At times, the internal partition walls of buildings have to be removed for some purpose or another, and if these are load-bearing walls a system of dead shoring has to be employed to support the floors and roof timbers whilst the work is being carried out. Fig. 8 illustrates a simple example of dead shoring which is used to support the floors and roof whilst the lower portion of the internal partition wall is removed and a lintel placed beneath the remaining portion of the wall. It will be noticed that 4 in. by 4 in. timbers have been used on each side of the partition wall to take the weight from the wall and transfer it down to surface concrete level.

The floor- and ceiling-joists are both supported by the wall and it is necessary to have 4 in. by 4 in. heads beneath the joists, supported by 4 in. by 4 in. posts at 4–5 ft. intervals across the widths of the rooms. The flooring at ground-level should be removed to allow the lower posts to go down to the surface concrete. Their ends should rest on plates so that the loads can be spread over a fairly large area. Folding wedges are used for easy adjustment. The top posts rest on the flooring in the upper floors, with a plate underneath their ends to spread the load over a large area as in the lower part of the building. The upper posts must, of course, be placed directly over the lower ones so that the loads will be transferred to the lower level correctly.

When the timbers have been placed and the folding wedges tightened, the upper portion of the partition wall will be supported by the ends of the first-floor joists thus enabling the lower portion of the wall to be removed.

If the joists of the first floor are not resting on the partition wall a different approach must be made to support the wall. The joists shown in Fig. 9 are running parallel with the wall. As the ends of the first-floor joists will not support the upper portion of the partition wall, 6 in. by 6 in. timber supports called needles are employed to do that work. Fig. 9 shows details of this type of support around the first-floor level. Holes large enough to take the needles have to be cut through the partition wall. The needles are passed through and supported as seen in the drawing. If possible the needles should rest up against the lower edges of the first-floor joists so that adequate support will be given to the posts above.

Fig. 10 shows the type of job this work would be employed to assist in carrying out, and shows that two needles supported by four 4 in. by 4 in. posts about 4 ft. apart should be capable of supporting the weight of the upper portion of the wall. The needles of course must be kept sufficiently high enough to get the lintel into position, and hardwood folding wedges should be used between the needle and lower position of the wall for obvious reasons.

Flying shores and further examples of dead shoring will be dealt with in an advanced volume.

This is the final chapter of "Practical Carpentry and Joinery" by A.B. Emary. "Advanced Carpentry and Joinery for Craftsmen" by the same author begins on the following page.

Shoring

When buildings are being repaired, or if structural alterations are being carried out, the walls of the building, and often the floors, have to be supported. Sometimes, too, walls have to be propped temporarily if they start to bulge. Raking shores have been covered in the companion volume *(Practical Carpentry and Joinery)*, but often complications arise so that the shoring has to be carried out to meet the particular conditions at the site.

Arched shoring. Fig. 1 represents the front and side elevations of a system of raking shores somewhat different from those already described. In such a case, perhaps a building on a main road must be supported, but the busy state of the roadway may induce the authorities to insist that no obstruction is placed in the way of passers-by. To avoid rakers straddling the pathway, a frame, made from 9 in. by 9 in. timbers, could form a kind of archway over the path.

The weakness in a shoring job such as this, of course, is the tendency for the top of the frame to be pushed outwards and on to the road. To prevent this a number of 2 in. stiffening boards can be bolted across the feet of the rakers and extended to the head and inside post of the frame. These stiffening boards will act as braces and tend to hold the frame in its correct shape and so overcome the outward force. The frame must be made to withstand any lateral movement parallel to the wall of the building; the side elevation shows how this can be achieved. Two 9 in. by 9 in. struts are cut to fit up against well-fixed cleats to the outside post and the sole piece. The latter should be about 10 ft. long and run along the curb to the pavement. The cleats, which should be at least 3 in. thick and 12 in. or more long, should be secured to the timbers with $\frac{3}{4}$ in. diameter coach screws. All the 9 in. by 9 in. timbers are butt-jointed to one another and the joints secured by large metal dogs.

Raker feet are cut to fit on top of the frame head and against a 3 in. cleat which also is fixed with coach screws. As the floor joists rest on the wall being supported, the centre lines of the rakers must go through to the centre of the support for the joists. The wall piece for this type of work should be 3 in. thick, and when prepared should first be secured to the wall using wall hooks (see *Practical Carpentry and Joinery*). For additional stability it should be allowed to extend down the head of the frame and fixed to this member.

The needles pass across the tops of the rakers, through the wall piece and into the brickwork, and should be prepared from 4 in. by 6 in. material. A few stiffening boards, in addition to those at the frame, should be bolted to the rakers and wall piece to give additional stiffness. Fig. 2 shows the details around the feet of the rakers and outside corners of the frame. As an alternative, the cleats at the tops of the struts can extend up to the top of the frame.

Flying shores. This type of shoring is much more efficient than raking shores. There is always a tendency for the tops of raking shores to be pushed outwards by the forces in the wall, but in flying, or horizontal shoring as it is sometimes called, the horizontal supports are placed parallel to those forces, and so offer greater resistance. The orthodox method of erecting flying shores is seen in Fig. 7.

This type of shoring may be used, for example, when a four-storey building in a terrace has had to be demolished. While preparations are made for another building to be erected in its place, the party walls of the adjacent houses must be given support to prevent damage. The obvious method would be to erect two horizontal shores to support the walls at the floors to the third and fourth storeys, and have rakers supporting the walls at the first floor level and the ceiling level of the top floor as shown in Fig. 7. If the building had three storeys, then one horizontal shore would be fixed at the same level as the lower horizontal shore in Fig. 7, with rakers from this to go down to the first floor level and up to the ceiling level of the second floor, see Fig. 3a. Centre lines are again important (see to the right of Fig. 7).

Process of erection. To erect the flying-shore system seen in Fig. 7, holes should first be made in the walls at the correct positions to receive the ends of the needles. Next the wall pieces should be prepared and fixed by means of wall hooks, and the needles and cleats then secured.

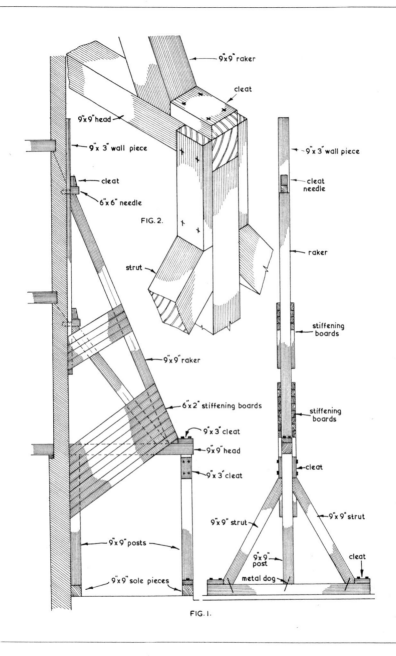

9"x9" raker

cleat

9"x9" head

9"x 3" wall piece

cleat

6"x 6" needle

FIG. 2.

strut

9"x 3" wall piece

cleat
needle

raker

stiffening
boards

9"x9" raker

6"x 2" stiffening boards

stiffening
boards

9"x 3" cleat

9"x 9" head

cleat

9"x 3" cleat

9"x 9" strut

9"x 9" strut

9"x 9" posts

9"x 9"
post

cleat

9"x 9" sole pieces

metal dog

FIG. 1.

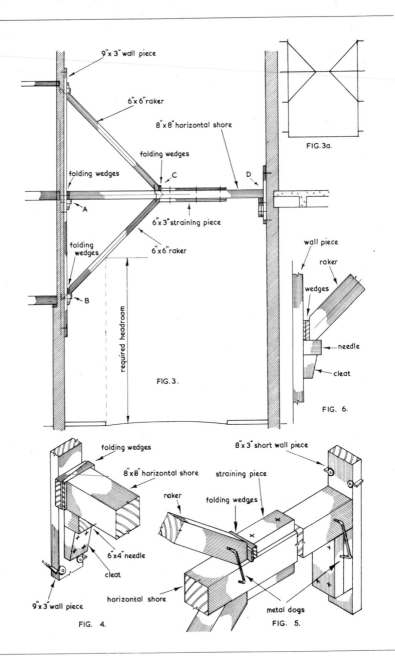

9″x 3″ wall piece

6″x 6″raker

8″x 8″ horizontal shore

folding wedges

folding wedges

C

D

A

6″x 3″ straining piece

folding wedges

6″x6″ raker

B

required headroom

FIG. 3.

FIG. 3a.

wall piece

raker

wedges

needle

cleat

FIG. 6.

folding wedges

8″x 8″ horizontal shore

8″x 3″ short wall piece

straining piece

raker

folding wedges

6″x 4″ needle

cleat

horizontal shore

metal dogs

9″x 3″wall piece

FIG. 4.

FIG. 5.

In a job of this size the cleats and needles should be prepared from 4 in. by 4 in. timber.

Next the two horizontal shores should be cut to length. These are 6 in. by 6 in. timbers offered up into position, and tightened by means of a pair of folding wedges at one end of each. The two vertical struts between the two horizontal shores should be positioned so that the rakers can be fixed top and bottom at an angle of approximately 45°. The struts can be cut from 4 in. by 3 in. material.

The four rakers are cut from 4 in. by 4 in. timbers and placed in position, the top two being tightened by a pair of folding wedges between the foot and the 4 in. by 3 in. straining piece nailed to the top of the horizontal shore. The lower two are tightened also by a pair of folding wedges placed at the foot of each as shown. A 4 in. by 3 in. straining piece is also fixed to the lower edge of the bottom horizontal shore to give the rakers an abutment. Joints should be securely fixed with large metal dogs.

Special conditions may exist for any kind of job of this sort. Fig. 3 for instance shows another system of flying shoring, where the timbers span the space across a narrow main thoroughfare, and where a certain headroom is required for passing traffic. As the wall of only one of the buildings needs support, it may be possible, if the building opposite is suitable and if permission has been obtained, to receive support from that building. It may then only be necessary to have one horizontal shore, as seen in the drawing. The building opposite has reinforced concrete floors, as shown, so it could be thought adequate to transfer all the forces from the building needing support to one point on the building opposite, in this case to a point immediately in front of a concrete floor. Only a short wall piece is needed for the building offering support. This can be fixed with wall hooks, and a double cleat to give a 4 in. to 6 in. seating for the horizontal shore is adequate at this point. Figs. 4, 5 and 6 show details at various points on the work.

Supporting an arcade. Occasionally, columns supporting arcading in Fig. 8, show signs of damage, and it may be necessary to dismantle the column completely and replace it with a new one. It is clearly necessary to support the arches immediately above the column whilst the work is being done. One method for carrying out this work is seen in Fig. 8. It consists of a yoke of two 6 in. by 6 in. timbers, each shaped on one edge to fit up against the surface of the arch with which it is in contact. The two pieces of the yoke are bolted together, as in the plan, and are supported at each end by a trestle made from 6 in. by 6 in. timbers.

The legs of the trestles rest on a sole piece with folding wedges between. The wedges are used for ensuring a tight fit for the yoke up against the arch surfaces, and if required could be placed between the yoke and top rail of the trestles instead of below the feet. To allow working space the trestles should be kept, say, 4 ft. apart. The trestle legs should preferably be splayed outwards to provide more space at floor level, but this is not absolutely necessary.

Dead shoring. This is necessary when a section of a wall of a building is removed for some purpose, such as renewing the beams over a shop front, enlarging an existing window in a building, and so on. Fig. 9 shows an elevation of a building which is having a large window incorporated in the ground floor external wall. A glance at the section through the building, Fig. 10, shows that the first and second floor joists are carried by the wall in which the alterations are taking place. It is first necessary to relieve the wall of these loads, and this is done by propping up the joists from a solid base with 6 in. by 4 in. timbers placed approximately 5 ft. apart across the front of the building. These props, of course, must have 6 in. by 3 in. heads and sole pieces, with folding wedges under each to tighten the work up efficiently.

The next decision is whether to give support to the outside surfaces of the external walls around where the work is to be carried out. If it is decided that this support is necessary, it can be in the form of raking shores as seen in the two drawings. It may also be considered that support should be given to the openings in the wall immediately above the alterations. In this case these are window openings, and the sides are supported to prevent damage to the brickwork. The sashes should be removed and the frames given extra strength by introducing side pieces and props to each opening. It may also be necessary to introduce props in an adjoining door opening as can be seen at ground floor level in Fig. 9.

The next step is to decide where the dead shoring should be positioned. They should be placed in the centres of the brickwork between the window openings, because the weight of the wall above the first floor openings is being transferred through these points.

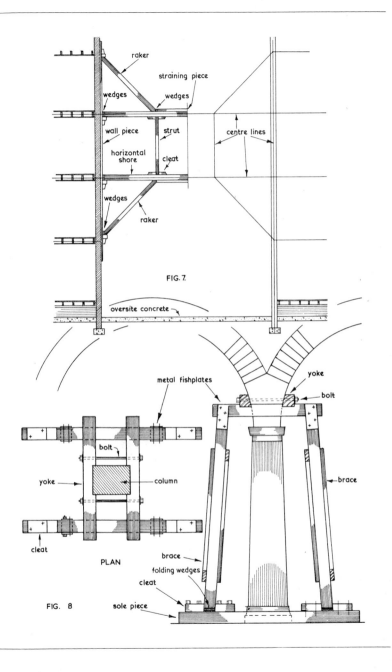

raker

straining piece

wedges

wedges

wall piece

strut

horizontal shore

cleat

wedges

raker

centre lines

FIG. 7.

oversite concrete

metal fishplates

yoke

bolt

bolt

yoke

column

brace

cleat

brace

PLAN

brace

folding wedges

cleat

FIG. 8

sole piece

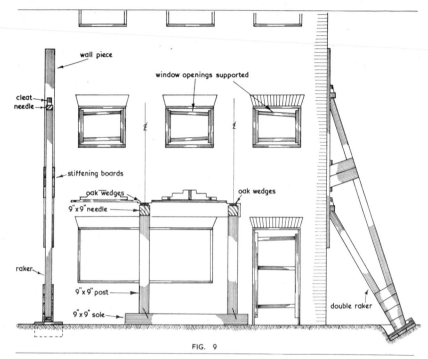

FIG. 9

When the positions of the needles of the dead shores have been decided, holes, a little larger than the needles, are made through the wall of the building. The 9 in. by 9 in. needles are then manhandled through these holes, and are supported at each end by 9 in. by 9 in. timbers, the lower ends of which rest on 9 in. by 9 in. sole pieces well bedded down on the ground. The joints between the timbers can be secured by large metal dogs. Large oak wedges are inserted in the brickwork holes and adjusted to take the weight of the brickwork immediately above the needles.

Only when it is confirmed that adequate support has been provided, and that all supports are doing their job, can work commence. The necessary brickwork on each side of the needles can be removed, as seen in the drawing, the new work introduced into the opening, and new brickwork incorporated to fill the space made by the removal of the original brickwork. Lastly the needles can be removed and the holes left filled with bricks to complete the work. Afterwards the floor supports and the rakers are removed.

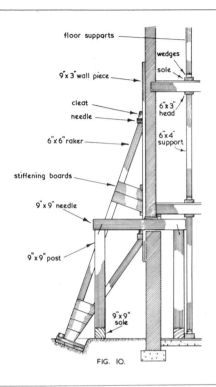

floor supports

wedges

sole

9"x 3" wall piece

cleat

needle

6"x 6" raker

stiffening boards

9" x 9" needle

9" x 9" post

6"x 3" head

6"x 4" support

9"x 9" sole

FIG. 10.

A view of raking shores. Note the stiffening boards and the cross bracing, both giving additional rigidity to the temporary supports.

19 Timbering to excavations

Many accidents, some serious, have been caused by thoughtlessness on someone's part when the sides of trenches and the like have collapsed and trapped workmen. When one realises that a cubic yard of soil can weigh up to a ton or more, it doesn't need much imagination to realise what the results would be if a person were walking down a trench 8 ft. to 10 ft. deep when the sides caved in. Risks should never be taken. It is easy to take the simplest way out when the ground around a trench looks firm and solid. 'We don't need any supports to that one' can be said all in good faith, but with serious, perhaps fatal, consequences.

When supports are needed. An excavation in the ground more than 3 ft. 6 in. deep should be supported. Fig. 1 shows the type of supports needed in firm hard ground which will remain in that state. Heavy lorries passing near by, continuous bad weather, and other conditions can cause havoc to what looks like safe ground. Poling boards, $1\frac{1}{2}$ in. thick, can be placed up to 6 ft. apart along the length of a trench, and these can be supported by 3 in. by 3 in. or larger struts. If the trench is fairly shallow, say up to 3 ft. 6 in. deep, only one strut to each pair of poling boards need be used.

Where firm hard gravel is being excavated, but there is a danger of the ground becoming less stable by reason of nearby traffic, etc., extra poling boards can be introduced as in Fig. 3. So as to keep good working space in the trench, waling pieces of 4 in. by 3 in. section can be introduced, and the struts kept at a distance of 6 ft. apart. Shallow trenches, again, can be supported by single struts, and to ensure that the struts keep their positions, the sides of the trench can be tapered slightly, as seen in Fig. 4.

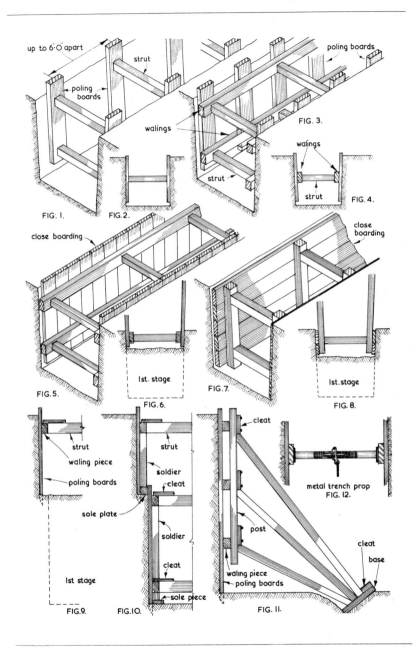

up to 6'·0" apart

poling boards

strut

poling boards

walings

FIG. 3.

FIG. 1.   FIG. 2.

walings

strut

walings

strut

FIG. 4.

close boarding

close boarding

1st. stage

FIG. 5.   FIG. 6.   FIG. 7.   1st. stage

FIG. 8.

strut

waling piece

poling boards

strut

soldier

cleat

sole plate

soldier

cleat

cleat

metal trench prop
FIG. 12.

post

1st stage

waling piece

poling boards

cleat

base

FIG. 9.   FIG. 10.   sole piece   FIG. 11.

197

Difficult ground. In loose and waterlogged ground, but where it is possible to dig a trench up to 3 ft. deep, close boarding as seen in Fig. 5 should be used. Waling pieces and struts as before are used for supporting the boards. It may be advisable to support the sides of the trench as the work proceeds, and this can be done as in Fig. 6. The poling boards can be introduced at an early stage and supported by one part of the walings and struts as shown.

In very loose ground, where it is difficult to dig the trench without the sides collapsing, the method shown in Fig. 7 should be used. This again is close boarding with the boards running in a horizontal direction. Fig. 8 shows how this can be done. Sufficient of the earth should be removed so that a board can be placed in position on each side and these supported by the soldiers and struts. As the depth increases so the next boards are introduced and the soldiers foreed down with a heavy hammer.

Deep trenches. Fig. 9 shows how a deep trench can be excavated and supported. The first stage can be made as in Fig. 6 until the required depth has been obtained. The lower ends of the poling boards are cut so that they can be forced into the soil by the heavy hammer. Waling pieces and struts will support the boards near the top of the trench. Soldiers can be placed below the walings to prevent these from slipping out of place. Their ends can rest on a sole piece.

The second stage can proceed similarly, the sides of this part of the trench being made somewhat closer to each other than those of the first stage. As before, waling pieces, struts and sole pieces should be used, and, to prevent any chance of the struts falling out of place, cleats should be firmly nailed to overlap the top surfaces of the walings.

Wide excavations. Fig. 11 shows how the sides of a wide excavation can be supported. Poling boards, 2 in. thick, can be placed against the sides of the excavation with three or more 6 in. by 4 in. waling pieces holding the boards against the vertical surface. Supports to these timbers are provided by 4 in. by 4 in. posts and rakers spaced at, say, 6 ft. to 8 ft. apart. Cleats should be bolted to the posts to give an abutment to the heads of the rakers.

Where practicable, metal props can be used in place of the timber struts, as seen in Fig. 12. These are easily adjusted and are more efficient than the struts.

20 Centres for Arches

Also see chapter 9.

Although arch work is fast disappearing in the building industry, it is still necessary for carpenters to be able to do this type of work because from time to time jobs of this nature do come along. The methods used for constructing the small arch centre were explained in the companion volume, *Practical Carpentry and Joinery*. Here we explain how large arch centres can be manufactured, and also the geometry and setting out of more difficult examples.

Semicircular arch. Fig. 1 is the elevation of a semicircular arch of 10 ft. span in a wall $2\frac{1}{2}$ bricks thick. On the left is shown how to overcome a projection at the springing line. The centre consists of two 7 in. by 2 in. ties, ribs from 7 in. by 2 in. material, 6 in. by 2 in. struts, and $\frac{3}{4}$ in. thick plywood gusset plates. The lagging which is nailed to the outside edges of the ribs is 2 in. by $1\frac{1}{2}$ in. All timbers are butt-jointed and the joints are secured with $\frac{3}{4}$ in. thick gusset plates with $\frac{1}{2}$ in. bolts and 2 in. square washers each end.

The centre is supported on three pairs of legs, each pair being secured together as a unit by means of 6 in. by $1\frac{1}{2}$ in. battens. Folding wedges are placed under the supports for final adjustment and for ease of dismantling. The supports should also be braced.

Fig. 2 is a vertical section through the centre and shows that the ribs are cross-braced. Note also that the length of the lagging pieces is about 1 in. less than the thickness of the wall, and the ribs are kept about $\frac{1}{2}$ in. from the ends of the lagging.

Semi-elliptical arch. Fig. 3 is the elevation of a semi-elliptical arch of 15 ft. span, and from front to back of any dimension. This type of archway could occur at the entrance of a building leading into a central quadrangle. Again all the timbers have been connected at the

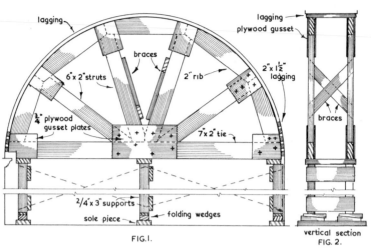

lagging

braces

6" x 2" struts

2" rib

2" x 1½" lagging

¾" plywood gusset plates

7" x 2" tie

2/4" x 3" supports

sole piece

folding wedges

FIG. I.

lagging

plywood gusset

braces

vertical section
FIG. 2.

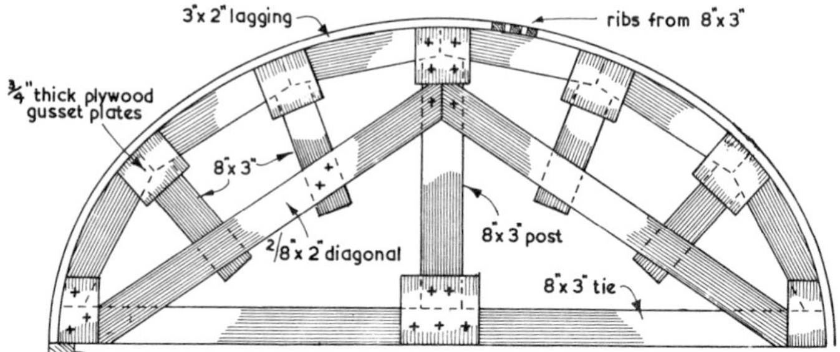

3" x 2" lagging

ribs from 8" x 3"

¾" thick plywood gusset plates

8" x 3"

2/8" x 2" diagonal

8" x 3" post

8" x 3" tie

FIG. 3.

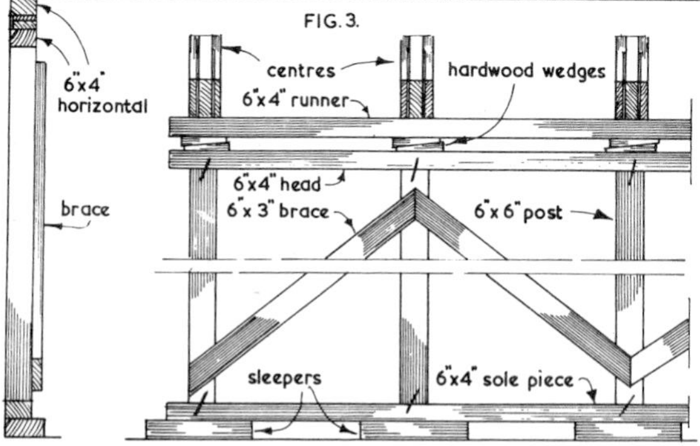

6" x 4" horizontal

brace

centres

hardwood wedges

6" x 4" runner

6" x 4" head

6" x 3" brace

6" x 6" post

sleepers

6" x 4" sole piece

FIG. 4.   ELEVATION OF SUPPORTS

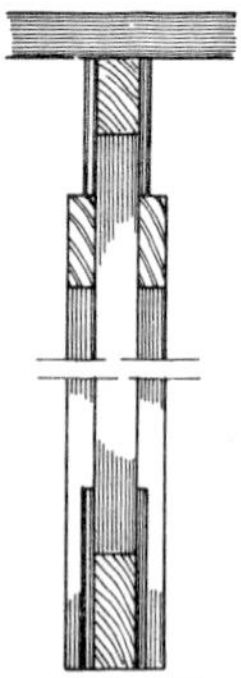

vertical section
FIG. 5.

intersections with $\frac{3}{4}$ in. thick gusset plates and bolts. Two pairs of diagonals support the joints round the ribs and the lagging is 3 in. by 2 in. The supports consist of a frame at each side of the opening, running from front to back. These frames are made from fairly heavy timbers and the joints are secured with metal dogs. The frames should be braced, see Fig. 4.

These supporting frames should be levelled, and can be placed on timbers such as old railway sleepers. A 6 in. by 4 in. runner is placed on the head of each frame, with hardwood folding wedges between these timbers and immediately over the posts of the frames. The centres are placed on the runners and over the frame posts which can be up to 5 ft. apart. The centres should be well braced through their centre posts. The wedges are used for final adjustment and dismantling.

Arch with splayed jambs. Fig. 6 shows the plan and elevation of an elliptical arch centre for an opening which has splayed jambs and a level crown. Given the shape of the ribs on the narrow side of the opening, the problem is to develop the shape of the ribs on its wide side. Draw the plan of the opening and place in the drawing the positions of the front and rear ribs. Also draw the elevation of the outline of the given rib. More details have been placed in this portion of the drawing than is actually necessary for geometrical purposes to show how the ribs will appear when made.

Divide the plan of the given rib into any number of equal parts, say eight, and project these points up to the curve in the elevation to give points 1–8. Project the edge of the jamb downwards to meet the centre line in X, and using this as a focal point, draw lines through points 1 to 8 in the plan to give points 1′–8′ on the edge of the required or rear rib. From these points draw vertical lines, and make these the same length as those on the elevation of the rib in elevation; for instance 8′–8″ should be the same length as 8–8 line, 7′–7″ the same as 7–7 line, and so on. A freehand curve through all these points will give an outline of the shape of the other rib.

Arch in curved wall. Fig. 8 is the plan and part elevation of a centre for a semicircular-headed opening in a curved wall. In other words it is a centre based on double curvature work. This drawing and the double curvature drawings in Chapter 19 should be compared. A

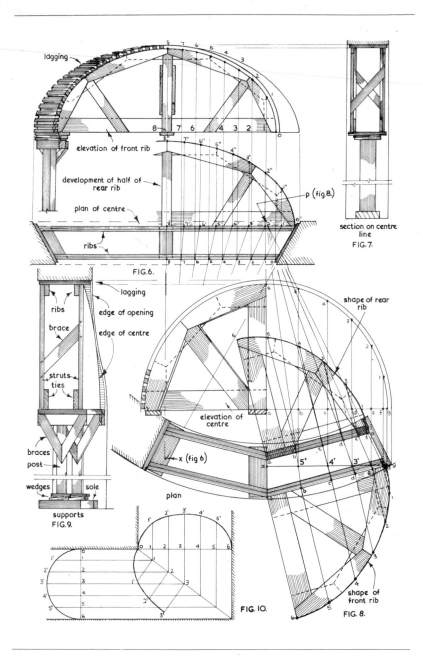

lagging

elevation of front rib

development of half of
rear rib

plan of centre

ribs

FIG. 6.

section on centre
line
FIG. 7.

shape of rear
rib

ribs
lagging
brace
edge of opening
edge of centre
struts
ties

elevation of
centre

x (fig 6)

braces
post
wedges   sole
supports
FIG. 9.

plan

p (fig.8.)

5'   4'   3'

shape of
front rib
FIG. 8.

FIG. 10.

glance at the plan shows that the wall is curved and that the jambs of the opening radiate back to point P.

The semicircular curve, which has been divided up into six equal parts, is the shape of the opening on line x–y in the plan. The positions of the ribs (the front and rear ribs are each made in two halves) are seen in the plan. To construct the shapes of the front and rear ribs, first divide one half of the elevation curve into, say, six equal parts, and project these points down vertically to the x–y line in plan, to give points 5′, 4′, 3′, etc. From these points draw lines to give points a, b, c, d, etc. on the front and rear ribs, using point p as a focal point.

Draw lines from points a, b, c, etc. on the front and rear ribs, perpendicular to the plan edges of each, and make the various perpendicular lines equal in length to those in the elevation. For instance, a–6 on the front and rear ribs should be equal in length to a–6 in the elevation. A curve through each set of points, 1, 2, 3, etc. will give an outline of the ribs required for the centre. A view of the ribs developed is shown in the drawings. Fig. 9 gives a vertical section through the centre.

Fig. 10 shows the geometry involved in setting out the shapes of the centres for the ceilings to passage-ways in the basement of a building. Given the semicircular shape of one, the other, which is wider, can be developed in the way shown.

21 Gantries

Gantries are used where building operations are taking place, and are erected over the pathway. They provide a loading and unloading platform for materials, and provide a base for scaffolding without obstructing the pavement immediately in front of the building. The platform enables the builder to store old brick rubble and other materials awaiting collection without causing inconvenience to passers-by.

Main frames. A gantry, Figs. 1 and 2, consists of two frames made from large timbers, such as 7 in. by 7 in. up to 12 in. by 12 in., butt-jointed together and secured with large metal dogs. One frame rests up against, or is very near to the building, and the other rests on the pavement by the curb. To keep the frames square, braces of, say, 6 in. by 3 in. material should be fixed by means of bolts or coach screws as shown in the elevation. The two frames should also be braced together across the width of the pathway as seen in Fig. 2.

Platform. The platform consists of 3 in. thick joists laid across the heads of the frames, and 2 in. or 3 in. thick boards to form the surface of the platform. As workmen will be carrying out certain work on the platform a safety rail should be provided round the three sides about 3 ft. 6 in. high to prevent possible accidents. If there is any risk of brick rubble or any other material falling down on to the pavement below, and this risk is always present if the platform is to be used for storing materials, close boarding should be fixed to the posts holding the safely rail. This boarding needs to be about 18 in. high. It may be an advantage to have one section of the safety rail with its close boarding easily removable so that the loading and unloading of lorries can be carried out readily.

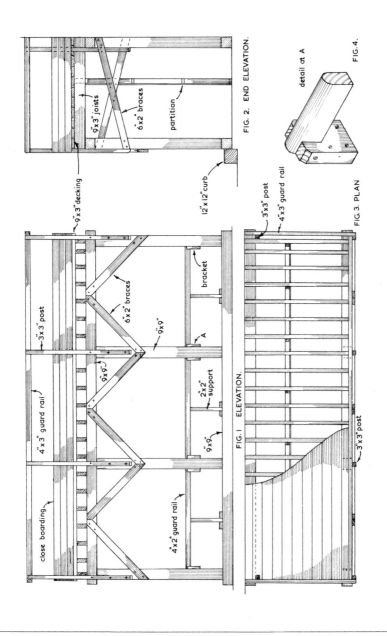

detail at A

FIG. 4.

9"x3" joists

6"x2" braces

partition

FIG. 2. END ELEVATION.

12"x12" curb

9"x3" decking

6"x2" braces

3"x3" post

bracket

9"x9"

9"x9"

A

2"x2" support

9"x9"

4"x3" guard rail

close boarding

4"x2" guard rail

FIG. I ELEVATION.

3"x3" post

4"x3" guard rail

FIG. 3. PLAN

3"x3" post

As the public will be passing underneath the gantry, a safety rail should also be provided between the uprights of the frame nearest the roadway. This will stop foot passengers unthinkingly stepping out into the roadway. The rail can be fixed and held in position with brackets, such as that seen in Fig. 4, and if a central support is required for each section, a 2 in. by 2 in. post, morticed and tenoned to the rail, can be provided as seen in Fig. 1.

It may also be necessary to narrow the width of the pavement below the gantry to provide a working space for work to be carried out at ground level. If this is required, a partition, as seen in Fig. 2, should be erected at the predetermined position. This consists of 3 in. by 2 in. timbers faced with $\frac{3}{8}$ in. external-grade plywood. Often it is also necessary to prevent vehicles on the roadway from accidentally damaging the structure by driving too close to the pavement. A large timber may be laid on the roadway, in the gutter immediately in front of the outer frame. If available, this timber, which is called a curb, should be either 9 in. by 9 in. or 12 in. by 12 in.

22 Formwork for Concrete

Also see chapter 8.

Reinforced concrete is being used more than ever at the present time, and in the foreseeable future its use will continue to increase. The carpenter who is not specially skilled in formwork, is sometimes called upon to design and erect the formwork for an odd job. It can be difficult for this man to imagine what the required 'box' or formwork will look like in relation to the finished piece of work. Fairly simple concrete formwork has already been covered in the companion volume, *Practical Carpentry and Joinery*, and it is the object of this chapter to go a little farther into the subject to help those not particularly skilled in this field.

Essential considerations. There are certain points to keep in mind when designing formwork for concrete.
1. The formwork is really a casting box into which a semi-fluid material is poured, and is shaped in accordance with the details obtained from the working drawings.
2. The boards or sheet material forming the box must be strong enough to avoid distortion or bending under the weight of the concrete.
3. The supports to the formwork must also be strong enough to support the formwork, the concrete, and possibly the weight of workmen and barrows.
4. The weight of the reinforcement must also be taken into consideration.
5. The striking of the formwork and its supports must be considered, too, so that the timber involved will not be unnecessarily damaged.

Materials. The timber used for shuttering is often some cheap form of softwood such as white deal. Resin-bonded plywood is being used widely for the formwork surfaces to floors, beams, columns, etc.

Metal formwork, too, is being used more extensively than it was twenty years ago. Although new systems are being introduced and the writer realizes how important these are to the industry, he thinks it is still necessary for the carpenter to be able to carry out this work in timber. If he can erect formwork in timber, he certainly should be able to carry out work with metal formwork systems if requested to do so.

Where a good smooth finish to the concrete is required, $\frac{1}{2}$ in. thick resin-bonded plywood sheets, adequately supported, should be used in place of boards. Alternatively, $\frac{3}{16}$ in. thick tempered hardboard can be used to line or surface the boards. Where boards only are to be used to achieve a fairly good surface the boards should be prepared on a planing machine. When a good surface is not necessary, rough boarding is used.

If possible, all formwork should be manufactured in the joiner's shop where machines are available, and if the formwork is to be used more than once this should be kept in mind so that the removal of the timber can be carried out with the minimum of damage. Often mineral oils are used to prevent the concrete from adhering to the timber. No formwork should be removed without the proper authority. Many accidents have occurred as a result of removing supports too soon.

Staircase formwork. Fig. 1 shows the plan of a concrete staircase of two flights connected by a half-space landing. Obviously, the first stage in this work is to provide the formwork and supports to the lower flight and the landing, Fig. 2.

Landing timbers should be erected first. These comprise $1\frac{1}{4}$ in. thick boards supported every 16 in. along their lengths by 3 in. by 2 in. joists. In turn the joists are supported by 6 in. by 3 in. runners and 4 in. by 3 in. posts. The posts rest on a sole piece with folding wedges between. These wedges will allow the timbers to be adjusted to the correct height and also allow the easy dismantling of the formwork when the time arrives.

The soffit of the first flight is of $1\frac{1}{4}$ in. boards, and these are supported every 16 in. on 3 in. by 2 in. joists. The last named are supported by three 4 in. by 2 in. runners, one near each side of the stairs and one in the centre. Posts, 4 in. by 3 in., positioned at right angles to the runners, transfer the weight of the concrete down to the sole piece on

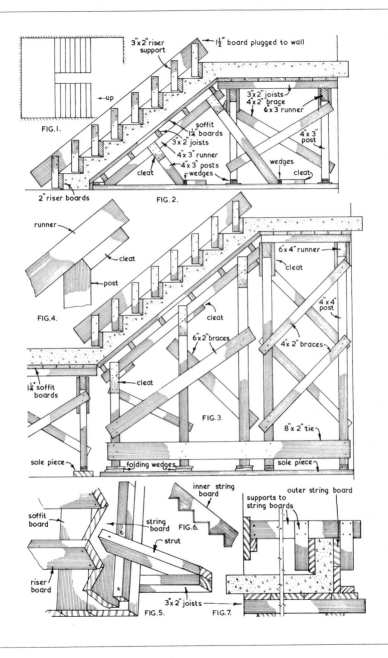

FIG.1.

3"x 2" riser support

1½" board plugged to wall

up

3"x 2" joists
4"x 2" brace
6"x 3" runner

4"x 3" post

soffit
1¼" boards
3"x 2" joists
4"x 3" runner
4"x 3" posts
wedges

cleat

wedges

cleat

2" riser boards

FIG. 2.

runner

cleat

post

FIG.4.

6"x 4" runner

cleat

4"x 4" post

cleat

6"x 2" braces

4"x 2" braces

cleat

1¼" soffit boards

FIG.3.

8"x 2" tie

sole piece

folding wedges

sole piece

soffit board

string board

inner string board

FIG.6.

outer string board

supports to string boards

strut

riser board

3"x 2" joists

FIG.5.

FIG.7.

which the posts rest. Cleats should be used for fixing the tops of the posts to the runners. Folding wedges are used at the foot of each post, and these are placed between the post and a cleat which is secured firmly to the sole piece. The latter should go back and rest against the wall for extra support, if possible.

Steps. The steps to the flight are formed by first plugging a board, to the wall, say $1\frac{1}{2}$ in. thick, to which the riser supports can be fixed. As the concrete of the stairs must pass into the wall to a depth of 4 in. to 6 in., the fixing of riser boards must be provided from above. Hence the need to plug the board to the wall. On the outside edge of the stairs, if the steps are to be similar to the cut string of a timber staircase, a 2 in. thick string board must be provided, see Fig. 5. This board, cut to the shapes of the steps, is secured to the top surfaces of the boards forming the soffit, and strutted from the ends of the 3 in. by 2 in. joists.

Riser boards $1\frac{1}{2}$ in. thick can then be fixed by screwing to the riser supports at the wall end and screwing to the vertical edges of the string board at the outer end, as in Figs. 2 and 5. It is usual to bevel the bottom edges of the riser boards to stop the edges from making a hollow mark in the tread of the steps, but as another finish to concrete is usually applied this is not important. Boards to form the tread or the horizontal surface to each step are not necessary.

The second stage of the work, the second flight of steps and the top landing, is shown in Fig. 3. This work is similar to the first flight and does not need a further description.

Posts are larger in section because of the greater lengths, and if these dimensioned timbers are not available, two pieces can be nailed together to make up the sizes. For instance, the 6 in. by 4 in. runners can be made from two pieces of 6 in. by 2 in. timber, two 4 in. by 2 in. pieces can be used for the posts, and so on.

If the staircase is to have a closed string on the outside, the balustrading can be formed as in Fig. 7. This consists of supporting an inner-cut string board as seen in Fig. 6, and an outer string board which will rest on the soffit boards. The top edges of the inner and outer string boards will run parallel with each other up the length of the flight. If metal baluster rails are to be fixed to the concrete balustrading at a later date it is possible that dovetailed blocks will

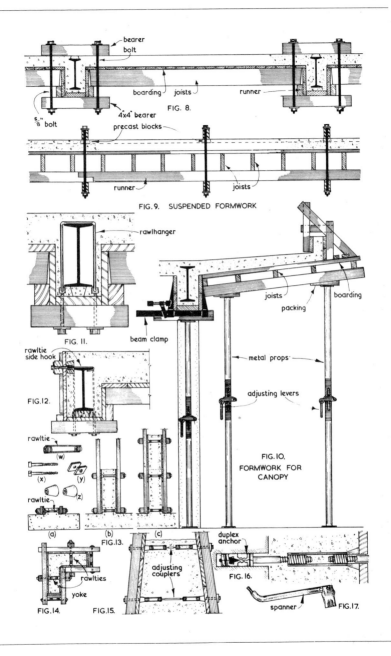

bearer
bolt

boarding   joists   runner

FIG. 8.

5/8" bolt

4x4" bearer

precast blocks

runner

joists

FIG.9. SUSPENDED FORMWORK

rawlhanger

joists   boarding

packing

FIG. 11.

beam clamp

metal props

rawltie
side hook

adjusting levers

FIG.12.

rawltie

(w)

(x)   (y)

rawltie

(z)

FIG. 10.
FORMWORK FOR
CANOPY

(a)   (b)   (c)

FIG.13.

duplex
anchor

adjusting
couplers

FIG. 16.

rawlties

yoke

spanner   FIG.17.

FIG.14.   FIG.15.

211

be required to be let in the top surface of the concrete so that the metal work can be grouted in.

Concrete floor. Figs. 8 and 9 are two longitudinal sections through a concrete floor with beams. The formwork is not supported from the floor below but suspended from the beams.

The 4 in. by 4 in. bearers are supported from above. Two bolts pass through the depth of the timbering to a bearer resting on the metal beam. The upper bearer is kept to the correct height above the beam by a precast concrete block. The bolts are covered with a sleeve of, say, cardboard so that they can be easily removed from the concrete when the timber is dismantled. The holes in the concrete are then filled.

Fig. 11 shows another method of supporting suspended shuttering. This is by the use of Rawlhangers. Fig. 12 illustrates how to form the side wall to a floor supported by suspended formwork. The main formwork is supported with Rawlhangers and the side shutter held in position with the use of Rawltie side hooks. These are similar to the Rawlhangers, but have a hook which passes round the flange of the beam. The dovetailed block is placed against the flange of the beam so that the bolt in the Rawltie side hook can be tightened up. Afterwards the block is removed and the hole in the concrete grouted in.

Concrete canopy. Fig. 10 is a view of the formwork and supports to a reinforced concrete canopy. The concrete forming the beam over the opening in the wall is formed by a three-sided box made to the required dimensions. Notice that the sides of the box go down to the lower surface of the bottom of the box. They are fixed by nailing through into the box bottom, thus strengthening considerably the timber used for the bottom. The beam sides can be supported by metal beam clamps as seen in the drawing, and the beam is held up at the correct height by using metal props. No wedges are necessary under the props because they can be adjusted in height by turning the adjusting levers which raise or lower the top section.

The soffit of the canopy is formed by using $1\frac{1}{4}$ in. to $1\frac{1}{2}$ in. boards supported by 3 in. by 2 in. joists spaced at 16 in. centres. These joists are carried on 4 in. by 3 in. runners at about 4 ft. centres which in turn, are supported on metal props. The outside upstand to the concrete canopy can be formed in the manner shown.

Concrete wall. Fig. 13a, b and c show a method of constructing a concrete wall with climbing formwork. Rawlties are used for these operations, and Fig. 13a shows the first step. To form the foundation to the wall a small upstand is fixed with Rawlties. The second stage, b, shows the wall taken up to the first lift, again using Rawlties and incorporating the ties that were used for forming the upstand. The next stage is seen in 13c. Formwork for the second stage has been raised so that more Rawlties are needed at the top of this next lift. Packing pieces are used to keep the bearers at the correct distance at the bottom.

A fourth stage, not shown, would entail the formwork being raised for the next layer of concrete to be poured, the top two Rawlties of the third stage being used as well as an additional one at the top of the new 'lift'. Fig. 13w, x, y and z show the various parts of a Rawltie. They consist of the welded section with the coiled steel sockets which form the thread for the bolts to be turned into; the two bolts, x; two large square washers, y; and z, the two hardwood cones. Only the welded portions are not recovered after use. Bolts, cones and washers can be used again with additional welded portions. Figs. 14 and 15 show how Rawlties can be used for formwork which would prove unsatisfactory if the welded parts were not available. Fig. 14 shows their use in a column of uncommon shape, and Fig. 15 illustrates how they can be used in conjunction with couplers. These last named can be used for adjusting the lengths of the fixture for a wall with tapered sides.

Fig. 16 illustrates how a Rawltie can be used with a Duplex anchor to fix formwork adjacent to an existing masonry wall. Fig. 17 shows the type of box spanner made for use with Rawlties, etc.

23 Roofs

Also see chapter 12.

Many advances have come about since the second world war in respect of roofing, due no doubt, to the cost of timber and the necessity of producing roofs of large span without intermediate supports with small section timber. These advances have covered the whole field of roofing from domestic buildings to factories. Roofs as far as domestic buildings are concerned tend to have flatter pitches than before, and many new trusses have been designed as a consequence, but, even so, the traditional pitched roof of 40° to 45° has not been forgotten in the attempt to conserve timber.

Perhaps the chief influence in these developments is the *Timber Research and Development Association* (TRADA) known originally as *The Timber Development Association* (TDA). This association has carried out research in many fields in which timber is the main material, and roofing has been one of their main subjects. The TDA truss (described in the companion volume, *Practical Carpentry and Joinery*), revolutionised roofing for domestic buildings, this truss making a far superior job compared with the old traditional method.

Many of the roof trusses described in the first part of this chapter have been designed by TRADA, and further information and drawings on many of these trusses and some others can be obtained from TRADA, Hughenden Valley, High Wycombe, Bucks.

Low-pitched roof (1). Fig. 1 shows the elevation of a low-pitched domestic truss, comprising principal rafters made from two pieces of 4 in. by $1\frac{1}{2}$ in., and a lower chord made also from two pieces of 4 in. by $1\frac{1}{2}$ in. Fig. 2 shows the joints at the intersection between the principal rafters and the lower chord. Three gusset plates are used, each 18 in. by 4 in. by 2 in., with three $\frac{1}{2}$ in. diameter bolts and four 2 in. double-toothed timber connectors to each bolt, see Fig. 2a.

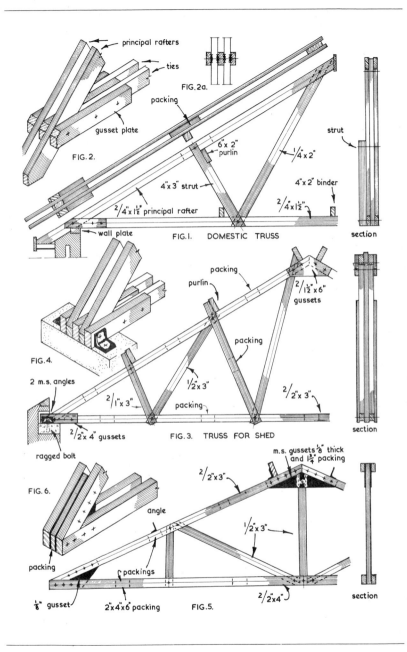

principal rafters

ties

gusset plate

FIG. 2.

FIG. 2a.

packing

6" x 2" purlin

1/4" x 2"

4" x 3" strut

4" x 2" binder

²/4" x 1½" principal rafter

²/4" x 1½"

wall plate

strut

section

FIG. 1. DOMESTIC TRUSS

packing

purlin

packing

²/1½" x 6" gussets

FIG. 4.

½" x 3"

2 m.s. angles

²/1" x 3"

²/2" x 3"

²/2" x 4" gussets

packing

ragged bolt

FIG. 3. TRUSS FOR SHED

section

m.s. gussets ⅛" thick and 1¾" packing

²/2" x 3"

FIG. 6.

angle

½" x 3"

packing

packings

⅛" gusset

2" x 4" x 6" packing

²/2" x 4"

FIG. 5.

section

One 4 in. by 2 in. tie is used in each half, and these are situated between the pairs of timbers forming the principal rafters and the lower chord. They extend down from the ridge to where the foot of the purlin support meets the lower chord.

Note that centre lines are important when setting out these trusses as they give the positions of the holes for the bolts which should pass through the centres of all the timbers. The purlins are 6 in. by 2 in., and their supports are 4 in. by 3 in. bolted on an outside surface of the principal rafter and lower chord.

When positioning on the roof, the principal rafter is notched over the wall plate, as seen in the drawing. The trusses should be placed approximately 6 ft. apart along the length of the roof, and binders, 4 in. by 2 in. in section should be used as required to prevent the sagging of the intermediate ceiling joists.

Large shed truss. Fig. 3 is the elevation of a truss for a large shed on which a covering such as corrugated iron or asbestos is to be used. That shown is suitable for a building with a span of about 25 ft. As the roof covering is much lighter than for the domestic truss, where it is presumed tiles will be used, the timbers for this truss can be somewhat smaller in size. For instance, the lower chord and the principal rafters each comprise two pieces 3 in. by 2 in., and are 2 in. apart to allow the single ties to pass between them. The other members in this case are made up from two pieces 3 in. by 1 in. and each is on an outside surface of the main timbers of the truss.

Fig. 4 shows the joint where the principal rafter and the lower chord meet. Assuming that the walls of the building are to be brick, a plate must be placed on the top of the wall at each end, and bevelled to the pitch of the roof so that the lower ends of the roof covering can be secured. In this case concrete slabs have been provided for the trusses, and they have been fixed to the masonry with two mild steel brackets. All joints are secured with bolts and timber connectors as before.

Low-pitched roof (2). Fig. 5 is the elevation of another low-pitched roof, designed to be used with roof sheeting such as corrugated asbestos, but where angle-iron purlin supports have been fixed to the top surfaces of the principal rafters. Metal gusset plates have also been used for joining the timbers at the eaves and ridge. The gusset

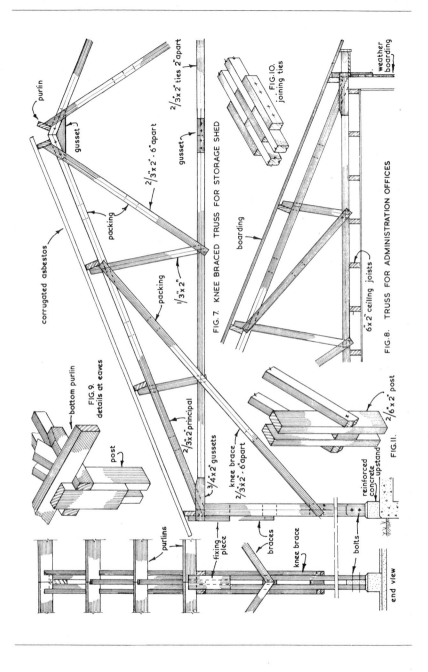

purlin

gusset

2/3"x 2" - 6" apart

packing

1"/3 x 2"

packing

2/3"x 2" ties 2" apart

gusset

FIG. 10.
joining ties

boarding

corrugated asbestos

FIG. 7. KNEE BRACED TRUSS FOR STORAGE SHED

6"x 2" ceiling joists

weather
boarding

FIG. 8. TRUSS FOR ADMINISTRATION OFFICES

bottom purlin

FIG. 9.
details at eaves

post

2/3"x 2" principal

3/4"x 2" gussets

knee brace

2/3"x 2" · 6" apart

2/6"x 2" post

FIG. II.

reinforced
concrete
upstand

purlins

fixing
piece

braces

knee brace

bolts

end view

217

plates ($\frac{1}{8}$ in. thick mild steel) are in pairs, and are situated between the principal rafters and the packing pieces.

Truss with knee braces. Fig. 7 is a timber truss suitable for either a storage shed, or for the roof over administration offices, or for some other similar purpose. When knee braces are incorporated in this, it is known as a knee-brace truss, and these components are almost essential when the storage shed is of the open variety. In other words, the roof might cover the space below but the sides of the shed could be completely open except for the upright supports to the trusses.

Let us assume that the trusses are going to be used for an open shed. The upright supports could rest on a concrete upstand as seen in the end view, and they could comprise two pieces 6 in. by 2 in. spaced 2 in. apart by packing pieces. The trusses, the end of one of which is seen in Fig. 9, rest on top of the supports and are secured with a fixing piece at the eaves. The trusses are made up of two 3 in. by 2 in. ties forming the lower cord. The principal rafters are each made from two 3 in. by 2 in. timbers. The knee braces, which are 6 in. apart, are bolted to the upright supports, the chords and the principal rafter as shown. Fig. 11 shows the joint at the feet of the knee brace. The trusses should be spaced 10 ft. apart and braces fixed between the vertical supports, as shown in the end view. Three 4 in. by 2 in. gusset plates are used at the eaves and one gusset plate at the ridge. The purlins are supported as shown.

The drawings also indicate where packing pieces, which considerably strengthen the structure, should be placed.

Fig. 10 shows how to build up the timbers for the lower chord or tie, where the required length of timber is not available. Gusset plates at these joints should be at least 24 in. long, and bolts and timber connectors used between each pair.

Fig. 8 shows how the truss can be used for office purposes. The knee braces are dispensed with and replaced with ties which extend down to the lower chord only. Ceiling joists 6 in. by 2 in. are fixed to the undersides of the lower chords, and these should be fixed at 16 in. to 18 in. centres according to the ceiling material being used.

The sides of the office building can be filled in between the truss supports with studding and weather-boarding to the outside, and plaster boarding to the inside surfaces.

Bow-string truss. The new form of bow-string truss is seen in Fig. 12.

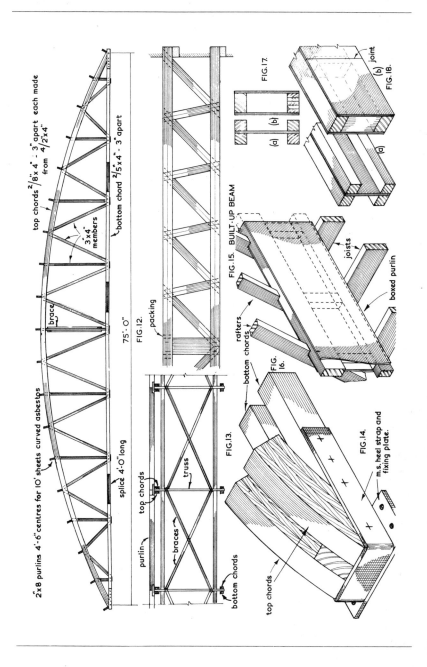

top chords $2/8" \times 4" - 3"$ apart each made from $4/2" \times 4"$

bottom chord $2/5" \times 4" - 3"$ apart

brace

3"×4" members

75'-0"

FIG. 12.

packing

splice 4'-0" long

2"×8 purlins 4'-6" centres for 10' sheets curved asbestos

FIG. 13.

purlin

top chords

truss

braces

bottom chords

FIG. 15. BUILT-UP BEAM

FIG. 17.

(a)
(b)

joint

(b)

FIG. 18.

(a)

rafters

bottom chords

joists

boxed purlin

FIG. 16.

FIG. 14.

bottom chords

top chords

m.s. heel strap and fixing plate.

219

Although this has been designed to span 75 ft., these trusses are capable of spanning a much greater distance, 150 ft. and more. Trusses comprise two top chords and two lower chords with purlin supports between. The top chords each have overall dimensions of 8 in. by 4 in., and are made up from four 4 in. by 2 in. timbers built up to the desired curve. Bottom chords can be of solid timber but can also be built up from smaller pieces in the same way as the top chord, if desired. Each bottom chord has an overall dimension of 5 in. by 4 in. The purlin supports are all 4 in. by 3 in. and are positioned as shown in Fig. 12. Lower chords should be given a camber of approximately 3 in.

Bow-string trusses are spaced along the length of the building at 12 ft. to 14 ft. centres and well braced, as seen in Fig. 13. A view of the joint at the eaves of a truss is seen in Fig. 14. A mild steel strap is made to pass round the outside surfaces of the heel, and a 3 in. packing piece placed between the chords will keep these the correct distance apart. The fixing plate is welded to the heel strap and used for securing the truss in position.

Truss for roof decking. Fig. 15 shows a view of a beam which could be used to carry roof decking. This consists of two upper and two lower chords with the pieces between positioned so that one bolt and three timber connectors are required for each intersection. Plywood box beams and plywood web beams, Figs. 16, 17 and 18, are also used a great deal for structural work in schools and buildings of this sort.

Open-type trusses. Figs. 19 and 22 illustrate open-type timber trusses, built on traditional lines. These trusses are found in public buildings such as church halls, churches, town halls and the like. They exemplify why so much research has gone into the design of roof trusses since the war. Large timbers, such as those shown in these trusses, are most uneconomical in present times.

The open-type truss, Fig. 19, is the type of roof truss seen supporting the roof to a village hall with a span of, say, up to 30 ft. Each truss consists of two principal rafters, a collar, and two wall pieces. The wall pieces, which help to prevent the feet of the principal rafters from spreading, are bolted to the walls of the building and rest on stone corbels. At the top, they are bridle-jointed to the lower ends of the principal rafters. To help restrict the tendency of the feet of the principal rafters to spread, mild steel straps are bolted to the

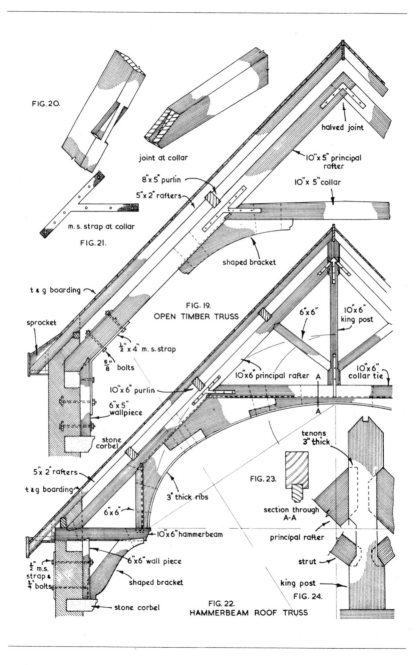

FIG. 20.

joint at collar

m. s. strap at collar

FIG. 21.

halved joint

10"x 5" principal rafter

10"x 5" collar

8"x 5" purlin

5"x 2" rafters

shaped bracket

FIG. 19.
OPEN TIMBER TRUSS

t & g boarding

sprocket

½"x 4" m. s. strap

⅝" bolts

10"x 6" purlin

6"x 5" wallpiece

stone corbel

5"x 2" rafters

t & g boarding

6"x 6"

6"x 6"

10"x 6" king post

10"x 6" principal rafter   A

10"x 6" collar tie

A

tenons 3" thick

FIG. 23.

section through A-A

principal rafter

3" thick ribs

10"x 6" hammerbeam

½" m.s. strap & ¼" bolts

6"x 6" wall piece

shaped bracket

strut

king post

FIG. 24.

stone corbel

FIG. 22.
HAMMERBEAM ROOF TRUSS

A view of some prefabricated timber trusses.

underneath sides of the principals and the fronts of the wall pieces.

A purlin is situated halfway between the wall plate supporting the common rafters and the ridge board, and the principal rafters are supported directly below the purlins by a collar. The collar is bridle-jointed to the principal, and mild steel straps are used, front and back, to strengthen these joints. Top ends of the principals are halved and, again, mild steel straps are used to secure this joint.

Shaped brackets can be used below the collar to give rigidity to the truss. Figs. 20 and 21 show the type of bridle joint to use at the ends of the collar and the purpose-made straps which are bolted to the timbers at these points.

Hammerbeam truss. Fig. 22 shows the elevation of a hammerbeam roof truss. These are used for supporting the roofs over buildings of up to 45 ft. span and consist of principal rafters, hammerbeams, wall pieces, collar, and shaped brackets. Depending on the design, some trusses also have king posts and struts, and occasionally, more than one hammerbeam to each principal rafter. The wall pieces are bolted to the wall, resting on stone corbels, and are bridle-jointed to the horizontal hammerbeam. A bracket rests on the same corbel as the wall piece and supports the other end of the hammerbeam.

The principal rafter is shown bridle-jointed to the hammerbeam over the wall of the building, and the centre of this member is supported by a collar. Shaped ribs are bolted to the underneath surfaces of the collar and principal rafters, and to the vertical member which rises from the inside end of the hammerbeam. These, as in the open-type roof, give rigidity to the truss.

Above the collar is a king post to which the top ends of the principals are secured with struts from the foot of this member, supporting the principals halfway between the collar and ridge. Notice that the truss

Roof trusses spaced at 2 ft centres, being fixed to the vertical framing. Note the plywood gusset plates at each joint. *Canadian Demonstration Homes*

has three purlins equally spaced along the principals, and assistance in supporting the principals at these points is given by the vertical piece at the lower end, the collar at the centre, and the strut at the top. Fig. 23 shows a section through the collar and shaped ribs, and Fig. 24 gives alternative methods for jointing members at the king post.

As both the trusses shown on page 41 are of the open variety, in other words all timbers of the roofs can be seen from the floor below, all the timbers in the roof should be prepared and should be good quality material. The roof is usually boarded with tongued-and-grooved boarding, face side downwards, and all timbers should be chamfered or moulded where possible.

24 Geometry and the Steel Square in Roofing

Despite what some craftsmen say regarding the use of the steel square in roofing, a good knowledge of geometry is still necessary to enable one to become efficient in the use of this valuable tool. The steel square is a geometrical instrument, and, so far as roofing is concerned, it is invaluable because, having obtained the span and pitch of the roof, it enables the lengths and the bevels of a roof to be ascertained. This also applies, of course, to the man who uses geometry and drawing instruments. Having obtained the span and the pitch, he should then be able to develop the lengths and bevels of all the parts which go to make the roof. As the geometry of roofs is basically what one has to use when applying the steel square, it is good policy to go over all the work involved in obtaining the lengths and bevels of all the members of an oblique-ended roof.

Roof with oblique end. Fig. 2 is the plan of a roof with an oblique end. Note that the jack rafters at the oblique end are all at right angles to the eaves. At the top of the page is a vertical section through the roof, and this is pitched at 45°. Below the section is the plan of the roof with the oblique end at the top, Fig. 1.

Common rafter length and bevels are obtained from the vertical section. Remember that this length e–b represents the length of the rafter down to the top outside edge of the wallplate. The overhang of the eaves has to be added on to this length, and half the thickness of the ridge board will have to be deducted, see Figs. 6 and 7. Now turn to the hip a–e at the lower end of the plan. To obtain its length construct a right angle at e and make e–g equal to the rise of the common rafters e′–e, see section. a–g is the length of the hips a–e and b–e. Half the thickness of the ridge board must be deducted from this' length, see Figs. 8 and 9. The overhang will have to be added to the length obtained. The plumb bevel or cut is seen at g and the seat cut at a.

To obtain the splay cut for the two hips at the square end, construct a right angle at a so that h is on the extended centre line of the roof. Then with compass point in a and radius a–g describe an arc to give g′ on the extended line a–e. The splay bevel is a–g′–h. The backing bevel for these two hips is found by drawing the line 1–3–2 across and at right angles to the plan of the hip. With compass point in 3 and pencil set just wide enough to touch the true length of the hip a–g, describe an arc to give point 4 on a–e. The backing bevel required is 3–4–2. The lengths and bevels for the short and long hip at the oblique end of the roof are found in exactly the same way as those for the hip just covered. For the remaining bevels we should now turn to Fig. 3.

Splay cuts. To obtain the splay cuts for the jack rafters (the plumb and seat cuts are the same as for the common rafters) one or more of the roof surfaces have to be developed. To obtain the splay cut for the jack rafters which intersect with the hips at the square end, either the end of the roof a–b–e can be developed, or one of the sides adjacent to the square end. Let us choose the surface e–b–c–f. To develop this surface place the compass point in e at the top of the vertical section, and with radius e–b describe an arc to give point b′ on the horizontal line brought out from e. Drop a vertical line from b′ to give points b′ and c′ on horizontal lines brought out from b and c in the plan. e–b′–c′–f is the developed surface and the splay cut is seen near point b′. This is obtained by drawing any line which is at right angles to the eaves to pass across the developed surface.

To obtain the splay cuts to the jack rafters at the oblique end of the roof the oblique end f–c–d must be developed.

Purlins. The next bevels to develop are those for the purlins. First draw a section of the purlin, any size, as seen in the vertical section, so that the extended edge 1–2 will intersect with the centre line in e′ (Fig. 3). Next, draw the plan of the purlin as seen in the roof plan. To obtain the cuts required the two top surfaces 1–2 and 1–3 have to be developed. Draw a horizontal line through point 1 in the section, and with centre 1 and radii 1–2 and 1–3 in turn, describe arcs to give point 2′ and 3′ on the horizontal line passing through 1. Drop vertical lines from points 2′ and 3′ to intersect with horizontal lines brought

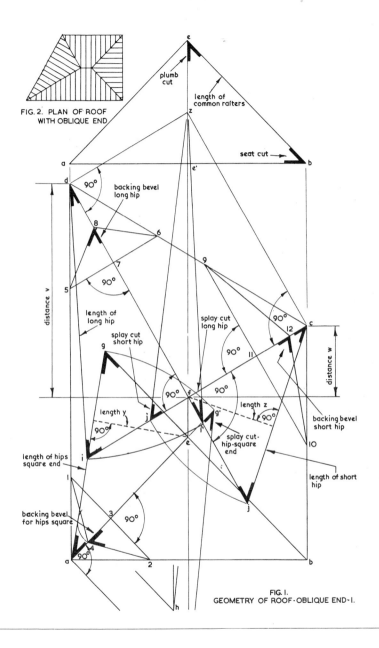

FIG. 2. PLAN OF ROOF
WITH OBLIQUE END

plumb
cut

length of
common rafters

seat cut

90°

backing bevel
long hip

distance v

length of
long hip

splay cut
long hip

splay cut
short hip

90°

90°

90°

90°

90°

90°

length y

length z

backing bevel
short hip

splay cut
hip-square
end

length of hips
square end

length of short
hip

backing bevel
for hips square

90°

90°

90°

distance w

FIG. I.
GEOMETRY OF ROOF-OBLIQUE END-I.

out from points 2 and 3 at each end of the purlin in the plan to give points 2′ and 3′ at each end. The developed surfaces are 1–2′–2′–1 and 1–3′–3′–1, and the bevels are seen at the ends of these surfaces.

Those near the bottom of the page are the bevels required for the purlins at the square end of the roof, and those near the top of the plan are for the purlins which intersect with the short hip. The purlin bevels to the long hip should be developed in a similar manner.

Lip bevels. The remaining bevels are the lip bevels to the purlins, and these are applied to the purlins in order for them to be fitted underneath the lower surfaces of the hip rafters, see Fig. 4. To obtain the lip bevels to the purlins which intersect with the hips at the square end extend the edge of the purlin 1–2 down to intersect with the point e′. Next draw the horizontal trace of the lower edge of hip e–b. This is drawn at right angles to the plan of the hip, and should extend from the centre line of the plan up to intersect with the horizontal line brought over from point e on the plan. From e″ draw a vertical line upwards to intersect with the base of the roof section in e‴.

From e‴ draw the vertical trace e‴–e. Extend the edge of the purlin 2–1 upwards until it meets the vertical trace in X, and with centre e′ and radius e′–x describe an arc to give x′ on the base line of the section. Drop a vertical line from x′ to give x″ on the e–e″ line, and from this point draw a line to x‴. The lip bevel is seen at x‴.

The lip bevel for the purlins which intersect with the small hip is found in exactly the same way. The horizontal trace f″–y‴ is drawn at right angles to the small hip, the vertical trace f‴–e giving point y. Radius e′–y will give point y′ on the base line of the section, and a vertical line dropped from y′ will give y″ on the f–f″ line. Draw line y″–y‴ for the bevel.

Steel square. The above is, briefly, the geometry needed for roofing work. Now let us turn to the steel square. Fig. 5 shows one side of a Stanley steel square, and the tables on the blade are those we should be interested in when dealing with roofing. These tables, of course, deal only with rectangular roofs. The top line of the tables deals with the length of common rafters per foot run. Let us just take one set of figures to illustrate the use of the tables, the figures under the 12 in. mark on the edge of the square.

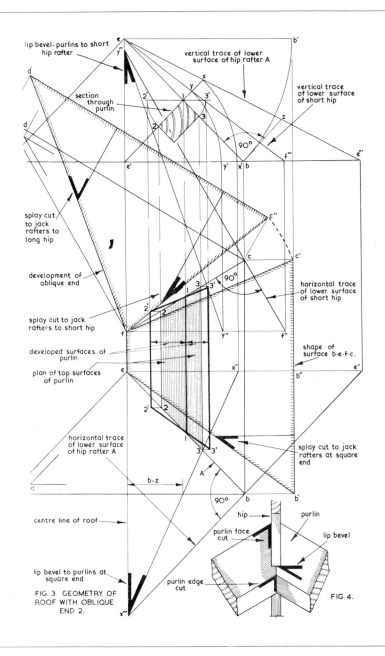

lip bevel - purlins to short hip rafter

vertical trace of lower surface of hip rafter A

vertical trace of lower surface of short hip

section through purlin

90°

splay cut to jack rafters to long hip

development of oblique end

splay cut to jack rafters to short hip

developed surfaces of purlin

plan of top surfaces of purlin

horizontal trace of lower surface of hip rafter A

horizontal trace of lower surface of short hip

90°

shape of surface b·e·f·c.

b-z

A

90°

centre line of roof

lip bevel to purlins at square end

splay cut to jack rafters at square end

FIG. 3. GEOMETRY OF ROOF WITH OBLIQUE END 2.

hip   purlin

purlin face cut   lip bevel

purlin edge cut

FIG. 4.

229

If we look at the top line of the tables the figures tell us that the common rafters are 16·97 in. long for every foot run when the rise is 12 in., see Fig. 6. If the rafter rise were 8 in. for every foot run, then the length of the common rafters would be 14·42 in. per foot run.

Let us assume that we have a roof to the dimensions of that seen in Fig. 6. Half the span is 12 ft. and the rise is 12 ft. The common rafter length is therefore 16·97 in. multiplied by twelve (the number of feet run) which is 16·97 ft., see Fig. 7. To this has to be added the overhang of the eaves, and half the thickness of the ridge board must be deducted from the length, see Figs. 7 and 12.

The lengths of the hip rafters are given in the second line of the tables. If we find that the length of the common rafters is 16·97 per foot run, we look at the second line and find that the hip rafters are 20·78 in. long for every foot run of the common rafters. If the common rafters have a run of 12 ft. then we multiply the figures 20·78 in. by twelve and find that the lengths of the hip rafters will be 20·78 ft., see Fig. 8.

To take another example, if the common rafters had a rise of 8 in. for every foot run, giving the length 14·42 in. per foot run, we would look under the dimension 8, go down to the second line of the tables, and find that the hips on that particular roof would be 18·76 in. long for every foot run of the common rafters. If the run of the common rafters were 12 ft. with a rise of 8 in. per foot run, the lengths of the hip rafters would be 18·76 ft. To this measurement, of course, must be added the overhang, and half the thickness of the ridge deducted, see Figs. 8 and 9.

There are two sets of figures for jack rafters, and these give the difference in the lengths of these members when used at 16 in. centres and 24 in. centres. It will be seen under the dimension 12 that the difference in the lengths of the jack rafters at 16 in. centres is $22\frac{5}{8}$ in. and the difference at 24 in. centres is $33\frac{15}{16}$ in., see Fig. 10.

The last two lines of the tables refer to the splay cuts for the jack rafters and hips and valleys.

Calculations without tables. If one prefers to work out all the lengths and bevels without referring to the tables on the blade, or if a roof with an oblique end such as that seen in Fig. 2 is to be constructed, the following notes and diagrams will help. They are useful to those who are not quite sure of the full use of the steel square in roofing. As

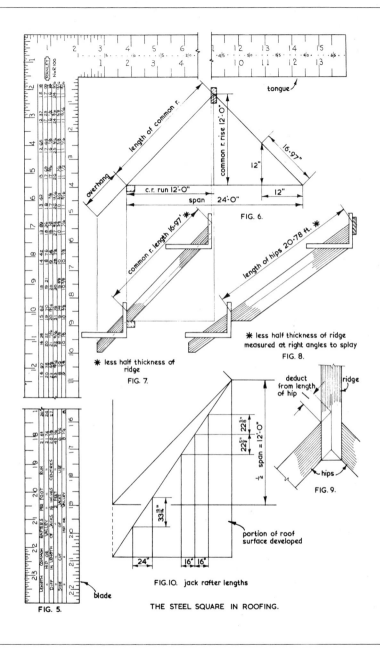

FIG. 6.

common r. length 16.97' **
** less half thickness of ridge
FIG. 7.

length of hips 20.78 ft. **
** less half thickness of ridge measured at right angles to splay
FIG. 8.

deduct from length of hip
ridge
hips
FIG. 9.

portion of roof surface developed

FIG.10. jack rafter lengths

FIG. 5.

THE STEEL SQUARE IN ROOFING.

the diagrams are explained it will be useful to refer to the geometry on the foregoing pages.

Fig. 11 shows how the lengths and bevels of the common rafters are found when the span and the pitch of the roof are known. Mark half the span on the blade and set the fence of the square up at an angle equal to the pitch of the roof to obtain the length and bevels. Remember that you are working to a scale of 1 in. to 1 ft. and so all the lengths will have to be multiplied by twelve. Also, half the thickness of the ridge must be deducted from the length obtained.

Next, the overhang of the eaves must be calculated, Figs. 12 and 13. Determine the distances a and b, and mark the total of these figures on the blade. Set the fence at the pitch and the length of the overhang is obtained.

The lengths of the hip rafters must be obtained in two steps. First, the run of the hips, square end. Mark half the span on the blade and half the span on the tongue. The distance across these points will give the run of the hips, square end. Fig. 14. The second step is to mark the run of the hips on the blade and the rise of the roof (or common rafters) on the tongue. The fence fixed across these points will give the lengths, plumb, and seat cuts for the hips, square end, Fig. 15. Remember the overhang of the hips must now be calculated and added to the length. Half the ridge thickness measured at right angles to the splay must be deducted.

To find the splay cut for the hips, square end, set the hip length on the blade and the hip run on the tongue. The bevel is found at the blade end of the fence. If reference is made to Fig. 1 it will be seen that the splay cut to the square end hips has been obtained by having the hip length as one side of the right-angled triangle, and the second side a–b is equal to the run of the hip. The bevel is found at g.

The backing for the hips, square end, is found in two stages. First, the distance y has to be found, Fig. 17. This distance in Fig. 1 is the amount we have to open the compasses so that they just touch the true length of the hip rafter. To find distance y set the run of the hips, square end, on the blade, and the rise of the roof on the tongue, and fix the fence over these two points. Measure with a two-foot rule the shortest distance between the corner of the square and the edge of the fence. This measurement is distance y. The second step is to set the fence to the run of the hip on the blade and distance y on the tongue. The backing bevel is at the tongue end of the fence, Fig. 18.

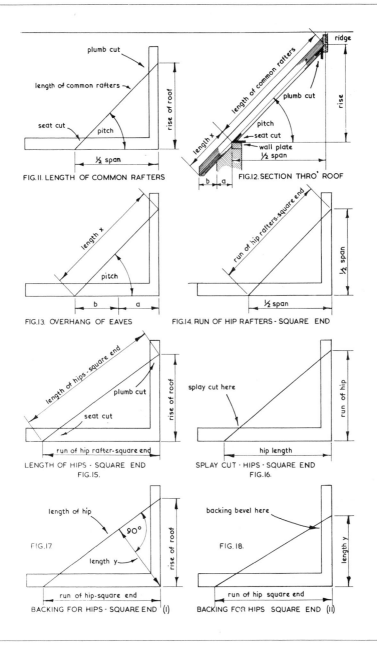

FIG.11. LENGTH OF COMMON RAFTERS

plumb cut
length of common rafters
seat cut
pitch
rise of roof
½ span

FIG.12. SECTION THRO' ROOF

length of common rafters
plumb cut
pitch
seat cut
wall plate
ridge
rise
½ span
length x
b a

FIG.13. OVERHANG OF EAVES

length x
pitch
b a

FIG.14. RUN OF HIP RAFTERS - SQUARE END

run of hip rafters-square end
½ span
½ span

LENGTH OF HIPS - SQUARE END
FIG.15.

length of hips - square end
plumb cut
seat cut
rise of roof
run of hip rafter-square end

SPLAY CUT - HIPS - SQUARE END
FIG.16.

splay cut here
run of hip
hip length

FIG.17
BACKING FOR HIPS - SQUARE END (I)

length of hip
90°
length y
rise of roof
run of hip-square end

FIG.18.
BACKING FOR HIPS SQUARE END (II)

backing bevel here
length y
run of hip square end

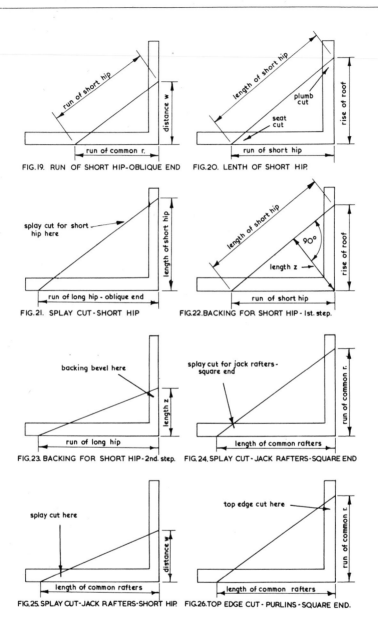

FIG. 19. RUN OF SHORT HIP-OBLIQUE END FIG. 20. LENTH OF SHORT HIP.

FIG. 21. SPLAY CUT - SHORT HIP FIG. 22. BACKING FOR SHORT HIP - Ist. step.

FIG. 23. BACKING FOR SHORT HIP - 2nd. step. FIG. 24. SPLAY CUT - JACK RAFTERS - SQUARE END

FIG. 25. SPLAY CUT - JACK RAFTERS - SHORT HIP FIG. 26. TOP EDGE CUT - PURLINS - SQUARE END.

We now come to the lengths and bevels for the hips at the oblique end. Let us take the short hip. To obtain the run of the short hip, mark half the span (or run of common rafter) on the blade, set the fence from this point at the same angle the small hip makes with the eaves at the square end of the roof (30°). Note the distance w as well as the run of the short hip, Fig. 19.

To obtain the length of the short hip, set off the run of the hip on the blade and the rise of the roof on the tongue. The fence will give the length of the hip as well as the plumb and seat cuts, Fig. 20. To obtain the splay cut for the short hip, set off the run of the long hip on the blade. This is equal in length to the line c–z, Fig. 1. Setting the length of the short hip on the tongue gives the splay cut at the tongue end of the fence, Fig. 21.

The backing for the short hip is found, as for that at the square end, in two stages. The first step is to obtain length z. This can be measured with the rule when the run of the short hip has been marked on the blade and the rise of the roof on the tongue. The second step is to mark the run of the long hip on the blade and length z on the tongue to obtain the backing bevel at the tongue end of the fence, Fig. 23.

Splay cuts. The splay cut for the jack rafters can be found simply by recalling what would be done if a surface of the roof had to be developed by geometrical means. The splay cut for all the jack rafters at the square end of the roof can be found by marking the length of the common rafters on the blade (this is equal to the width of the surfaces) and the run of the common rafters on the tongue (this is equal to b′–b′′ on the plan). The splay cut is found at the blade end of the fence, Fig. 24.

The splay cut for the jack rafters adjacent to the short hip is found by marking the length of the common rafters on the blade (the width of the surface) and distance w on the tongue to get the bevel at the blade end of the fence, Fig. 25.

Purlin cuts. Obtaining the cuts for the purlins is similar to developing the surfaces of a roof. The top surface of a purlin can be taken as a small portion of the roof surface which it supports. Let us consider the purlins at the square end of the roof first. It will be seen from Fig. 3 that the top surface is a portion of the roof surface. Consequently, when developed the end of the purlin will be of similar shape as the

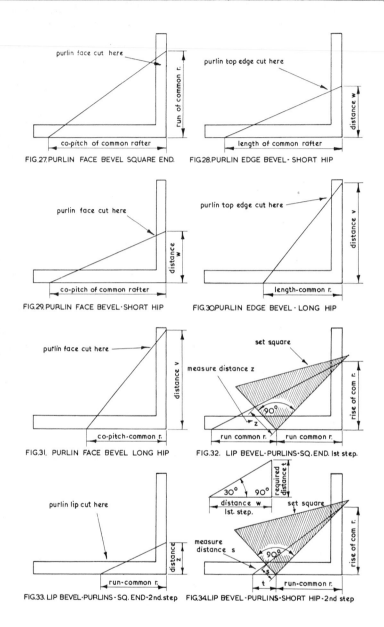

FIG.27. PURLIN FACE BEVEL SQUARE END.

FIG.28. PURLIN EDGE BEVEL - SHORT HIP

FIG.29. PURLIN FACE BEVEL - SHORT HIP

FIG.30. PURLIN EDGE BEVEL - LONG HIP

FIG.31. PURLIN FACE BEVEL LONG HIP

FIG.32. LIP BEVEL - PURLINS - SQ. END. lst step.

FIG.33. LIP BEVEL - PURLINS - SQ. END - 2nd. step

FIG.34. LIP BEVEL - PURLINS - SHORT HIP - 2nd step

end of the roof surface. Fig. 26 should make it quite clear that when developing the splay cut for the jack rafters the top edge bevel for the purlin can be obtained from the other end of the fence. When dealing with roofs pitched at 45° the side bevel for the purlin is exactly the same as for the top edge where the roof is rectangular.

This however only applies to roofs of 45°. If the pitch is more or less than 45° the bevel for the side is different from the top edge. So as not to make a common error, it is always best to follow a rule when developing the bevels to purlins. The length of the common rafter should be used when developing the top edge cut of the purlin, and the co-pitch (see Fig. 36 a and b) used when developing the side cut. To develop the top edge cut for the purlins adjacent to the short hip at the oblique end, mark off the length of the common rafter on the blade and distance w on the tongue (Fig. 1). The bevel required is at the tongue end of the fence, Fig. 28.

To obtain the face bevel to the purlins at the short hip, mark the co-pitch on the blade and the distance w on the tongue to obtain the bevel at the tongue end of the fence, Fig. 29.

Hips. Now let us turn to the long hip at the oblique end. Mark the length of the common rafter on the blade and the distance v, Fig. 1, on the tongue. The top edge bevel is found at the tongue end of the fence, Fig. 30. The face bevel to the long hip is found by marking the co-pitch of the rafters on the blade and the distance v on the tongue to obtain the bevel at the tongue end of the fence, Fig. 31.

Co-pitch. One might well ask, 'how does one find the co-pitch of a roof?' Let us take a look at Fig. 36b. The pitch of the roof is 30°, which means the co-pitch angle is 60°. The co-pitch is always at right angles to the slope of the roof surfaces. To find the co-pitch of the 30° roof, Fig. 36b, mark off half the span on the blade, set the fence up to 60° from the half-span mark, and measure with a rule the distance along the fence giving the length of the co-pitch.

Purlin lip bevels. The only remaining cuts required are the purlin lip bevels. Study Figs. 3 and 32 together. The horizontal trace of the hip b–e terminates at e″. The line e–e″ is equal to twice the run of the common rafters. Mark this distance on the blade. The vertical trace of the same hip is e‴–e. The distance between e and e′ is equal to the rise of the roof. Mark the rise on the tongue and fix

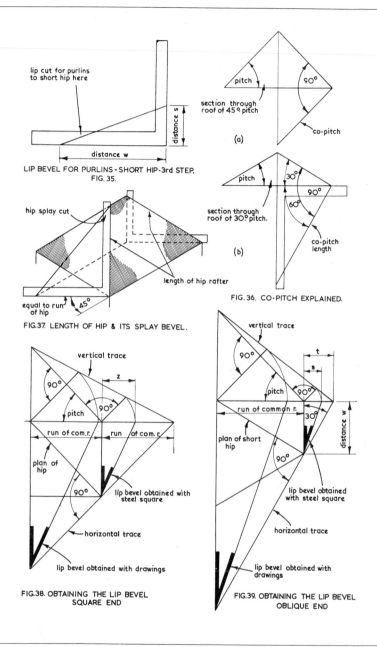

lip cut for purlins to short hip here

distance s

distance w

LIP BEVEL FOR PURLINS - SHORT HIP-3rd STEP.
FIG. 35.

pitch

90°

section through roof of 45° pitch

co-pitch

(a)

pitch

30°

90°

60°

section through roof of 30° pitch.

co-pitch length

(b)

FIG. 36. CO-PITCH EXPLAINED.

hip splay cut

length of hip rafter

equal to run of hip

45°

FIG.37. LENGTH OF HIP & ITS SPLAY BEVEL.

vertical trace

90°

z

pitch

90°

run of com. r.   run of com. r.

plan of hip

90°

lip bevel obtained with steel square

horizontal trace

lip bevel obtained with drawings

FIG.38. OBTAINING THE LIP BEVEL
SQUARE END

vertical trace

90°

t

s

pitch   90°

run of common r.

30°

plan of short hip

90°

distance w

lip bevel obtained with steel square

horizontal trace

lip bevel obtained with drawings

FIG.39. OBTAINING THE LIP BEVEL
OBLIQUE END

the fence over these two points. Now place a set square over the steel square as seen in the drawing, and measure distance z. This is equal to the distance b–z in the section and b–z in the plan. The next step in obtaining the lip bevel for the square end purlins is to mark off the run of the common rafters on the blade and distance z on the tongue. The lip bevel is at the blade end of the fence, Fig. 33 (see also the plan of the roof Fig. 4).

Obtaining the lip bevel for the short hip is illustrated in Fig. 34 (first and second steps). Fig. 35 (third step) is explained in geometrical form in Fig. 39. The reader should study these drawings and satisfy himself that the bevel obtained is the correct one.

Fig. 38 further illustrates how the lip bevel for the square end purlins is obtained, and Fig. 37 shows how the steel square is used for developing the splay cuts for the hip rafters.

25 Doors

Also see chapter 15.

Work involving doors will always play a big part in the life of the carpenter and joiner. The common types of door were dealt with in the companion volume *(Practical Carpentry and Joinery)* but there remain several which must be included in a book on advanced work.

Double-margin door. Where a single door has to be fitted to a rather wide opening, a double-margin door is often used, Fig. 1. In appearance it looks like a pair of doors. On the right of the drawing is shown the method used for constructing and assembling it. As can be seen, four stiles are used, the middle two being secured together by means of pairs of folding wedges.

The two halves should first be constructed as though they were to be used as double doors. Mortices are made across the width of the meeting stiles in line with each other. The doors are first assembled without their panels as separate units, glue being used on the mortices and tenons to the centre stiles only. The two halves are brought together and secured with the folding wedges through the centre stiles. When the ends of the wedges have been trimmed, the outside stiles are removed, the panels inserted into their grooves, the outer stiles glued, and the whole reassembled. The drawing on the right shows the panels being assembled into the nearly completed door. Fig. 2 is a cross section through the meeting stiles.

Vestibule screen. Fig. 5 shows a vestibule screen with fixed side lights and double swing doors. These are situated just inside the main entrance to a building, such as a large store, show rooms, hotels, etc. The screen shuts off the main part of the building from the outside. The elevation shows that the screen is fully glazed and has a transom.

Fig. 6 is a horizontal section through the screen, and Figs. 7 and 8

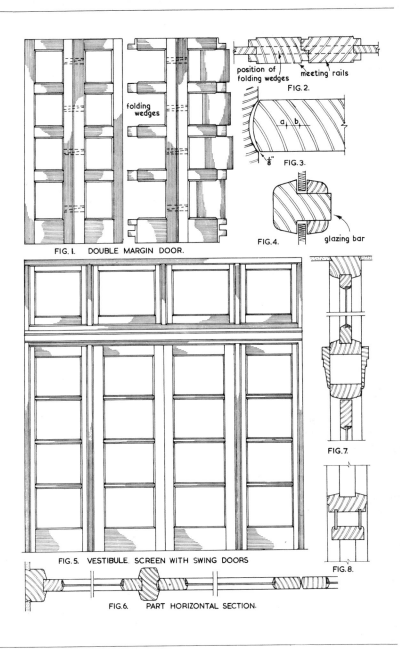

position of folding wedges

meeting rails

FIG. 2.

folding wedges

a b

8"

FIG. 3.

glazing bar

FIG. 4.

FIG. I. DOUBLE MARGIN DOOR.

FIG. 7.

FIG. 5. VESTIBULE SCREEN WITH SWING DOORS

FIG. 8.

FIG. 6. PART HORIZONTAL SECTION.

show two methods of building up the transom. The side lights and the fanlights above the transom are all fixed and are incorporated in the frame when this is being assembled.

Fig. 3 gives a section through one of the hanging stiles of a door, and a method for determining the curves of the frame and stile. Point b is the centre of the double-action floor hinge, and is used for obtaining the shape of the edge of the door stile. Point a, which is approximately $\frac{1}{2}$ in. away from b, is used for obtaining the shape of the hollow in the door frame. About $\frac{1}{8}$ in. joint is required at the edge of the door, and if the curves are set out as shown it is impossible to see through the joint. Fig. 4 is a section through a glazing bar with glazing beads. The beads should be on the sides of the screen facing the inside of the building.

Traditional type door. Fig. 9 is the elevation of another internal door, constructed in a traditional style, and is the type found in better class work. It is a five-panelled door with the inclusion of a frieze or intermediate rail, which is immediately below the top rail. There are many ways of preparing the framework and panels of a door, and Fig. 10 shows several of these. At a the door framework is moulded on each side, and the panels are raised and fielded on each side. The linings of the door are also framed and the panels to the linings are fielded and raised to match the door. The linings and built-up architrave are fixed to grounds. The architrave extends down to a plinth block, Figs. 9 and 11, to which the architrave and the skirting are scribed.

The framework at b has its framing left square-edged with its panels raised and fielded on one side only. Bolection mouldings are fixed to the framing around the panels on the raised side, and planted mouldings on the rear side. The methods used for fixing these mouldings were described in the companion volume, *Practical Carpentry and Joinery*.

At c is shown a similar finish but the planted moulding is replaced by a stuck moulding. Screws would not be used for fixing the bolection moulding in a case such as this, oval brads or lost-head nails being used, their heads being carefully punched below the surface and the holes filled.

At d are details of a door with double-raised and fielded panels, and framed double bolection mouldings. The bolection mouldings and panels would be made up as single units, mitred, tongued, and

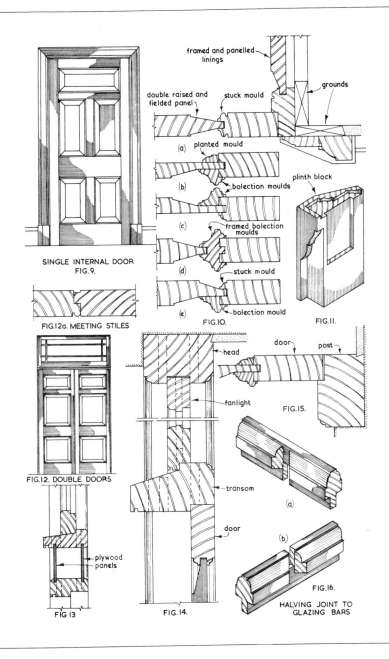

framed and panelled linings

grounds

double raised and fielded panel

stuck mould

(a) planted mould

(b) bolection moulds

(c) framed bolection moulds

(d)

stuck mould

(e) bolection mould

FIG.10.

SINGLE INTERNAL DOOR
FIG. 9.

FIG.12a. MEETING STILES

plinth block

FIG.11.

FIG.12. DOUBLE DOORS

plywood panels

FIG 13

head

fanlight

transom

door

door

post

FIG.15.

FIG.14.

(a)

(b)

FIG.16.

HALVING JOINT TO
GLAZING BARS

screwed at their corners, and included in the door framing during assembly.

At e the door has double-fielded and raised panels, the sides not matching. One side has stuck mouldings worked on the edges of the frame, and the other side has bolection mouldings. There are many variations to those shown in Fig. 10.

External doors. Fig. 12 shows the elevation of double external doors with a transom and fixed fanlight above. Fig. 13 gives another method of building up a deep transom, and 12a is a section through the meeting stiles of the doors. Each door has three panels and a frieze rail. The panels are raised and fielded and the framework has stuck moulding worked on each side.

Fig. 14 is an enlarged view of the details around the fanlight, and shows that, although the top rail (and stiles) of the fanlight fit into recesses in the door frame, its bottom rail is rebated to fit over the edge of the weathered rebate of the transom. The bottom edge of the rail has a groove worked into it to prevent water passing through to its back edge. If the bottom rail of the fanlight were to be fitted into the top surface of the transom, trouble from moisture would probably result. Fig. 15 is a section through a post of the door frame and a stile of one of the doors.

Often it is required to make a halving joint at the intersections of the glazing bars, as in the fanlight, Fig. 12. This halving joint is prepared as shown in Fig. 16a and b. If required, the mouldings can be scribed instead of mitred.

Doors with elliptical fanlight. Fig. 17 shows the elevation of a pair of external doors, with a transom and an elliptical-headed fanlight. In such a case it is much better to keep the transom at least 2 in. below the springing line so as to allow the joints between the door post and transom and the door post and head to be prepared without affecting the strength of each. Fig. 18 shows the positions of these joints. It can be seen that the post of the door frame extends to well above the top edge of the transom, keeping the two joints well apart. Fig. 19 gives the types of joints used at this point.

Figs. 20 and 21 give two ways of finishing the middle rails of the doors. The first, Fig. 20, has square shoulders to the rail which is made up in two halves so that the circular panel can be assembled. In the second method, Fig. 21, the rail has splayed shoulders as well as

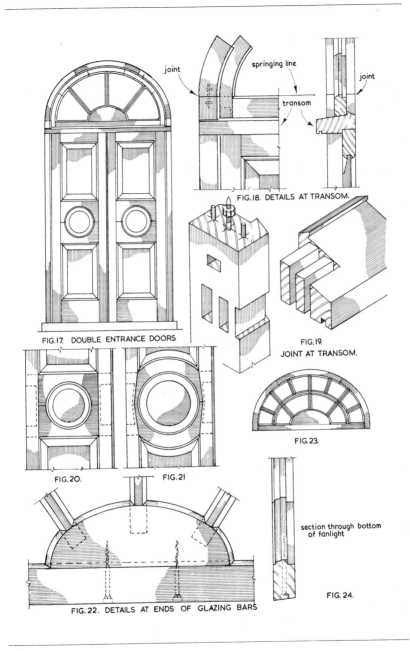

joint

springing line

transom

joint

FIG. 18. DETAILS AT TRANSOM.

FIG. 17. DOUBLE ENTRANCE DOORS

FIG. 19.
JOINT AT TRANSOM.

FIG. 23.

FIG. 20.

FIG. 21

section through bottom
of fanlight

FIG. 24.

FIG. 22. DETAILS AT ENDS OF GLAZING BARS

245

curved top and bottom edges, allowing for a larger circular panel.

Fig. 22 shows the bottom of the fanlight. To prevent complicated joints at this point, and to improve on the look of the work, a shaped piece is fixed to the top surface of the bottom rail, into which the lower ends of the bars are morticed. Fig. 24 is a side view of the shaped piece, and Fig. 23 shows another fanlight design.

Pointed-arch doors. Church doorways usually follow a shape based on pointed-arch design. Fig. 25 is the elevation of a pair of church doors based on the methods used for framed, ledged and braced doors, see the companion volume, *Practical Carpentry and Joinery*. Fig. 26 gives a horizontal section through the doors. To the left is an outside elevation and on the right is an inside view showing cross bracing which assists in keeping the doors square. All the inside edges of the frames are stop-chamfered, and each door is cross-braced as shown.

The joints of the head of the door frame can be secured with hand-rail bolts and dowels, and the curved portion of the doors joined at the springing line by using the joints shown in Fig. 27. At the top the usual mortice and tenon joint is used.

Tudor arch door. Fig. 29 shows the top part of a door which has to fit into an opening with a Tudor arch. These joints can be secured with handrail bolts or hammer-head key joints as shown in Fig. 30a and b. The method employed for setting out a Tudor arch is shown in Fig. 28. Let b–d be the width of the opening, and a–c the rise. Draw the rectangle a–3–b–c and divide b–3 into three equal parts. Join 2 to a, and with centre b and radius b–2 describe an arc to give 2′ on b–c. Make a–2″ equal b–2′ and bisect 2′–2″. Make 2–a–x a right angle, and extend a- x downwards until it meets the bisecting line in x. Draw a line from x through 2′. With centre 2′ and radius 2′–b describe the first part of the curve from b round to the line x–2′ to give point z. With centre x and radius x–a describe the other part of the curve a–z. The other half of the curve is found in exactly the same way as the first half.

Flame-resisting doors. Occasionally doors are required which are capable of resisting flames for a period of at least 30 minutes. There are several ways of making them, and two are seen in Figs. 31 and 33. Fig. 31 is the elevation of a panelled door, and the section through the door, Fig. 32, indicates that the panels are the same thickness as the

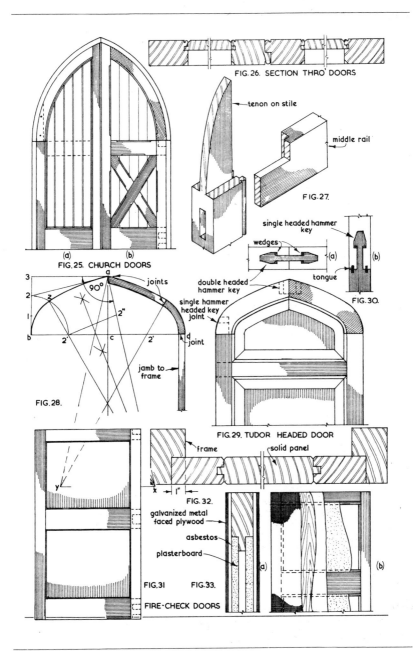

FIG. 26. SECTION THRO' DOORS

tenon on stile

middle rail

FIG. 27.

single headed hammer key

wedges

double headed hammer key

tongue

(a)   (b)

FIG. 30.

(a)   (b)

FIG. 25. CHURCH DOORS

3
2
1
b
a
90°
joints
2"
2'   c   2'
d
joint

joints

single hammer headed key joint

jamb to frame

FIG. 28.

FIG. 29. TUDOR HEADED DOOR

frame

solid panel

x → 1"

FIG. 32.

galvanized metal faced plywood

asbestos

plasterboard

y

(a)   (b)

FIG. 31   FIG. 33.

FIRE-CHECK DOORS

framework. The door should be made from a timber which has high natural fire resistance such as jarrah, karri, padauk, teak, gurjun, and greenheart. Many other and more common timbers are resistant to fire but to a lesser degree, such as ash, beech, idigbo, iroko, English oak, sycamore, and so on. The framework and panels of the door should be at least 2 in. thick throughout and the depth of the rebate in the door frame should also be 1 in., see Fig. 32.

Another method used for constructing doors with a good fire resistance is illustrated in Fig. 33. This consists of making the frame of the door so that panels of plaster board can be inserted flush to the two surfaces of the frame. Sheets of asbestos and metal-faced plywood are then secured to both faces, as seen in the section, Fig. 33.

A view of a church doorway showing inside and outside details of the doors.

26 Sliding Door Gear

Room doors. Figs. 1 to 6 show details of the *Marathon* ball-race gear suitable for doors in houses and flats up to a weight of 150 lb. It is designed for internal doors but can, if required, be used externally if properly protected. The sets as purchased from the manufacturers consist of the complete top assembly, a bottom channel for fixing to the bottom surface of the door, a floor guide, and one end or centre stop. If special fittings are required the manufacturers will supply them.

Fig. 1 illustrates the common method used for fixing the *Marathon* gear. The top assembly is screwed to a board which is fixed above the door opening, and a metal pelmet is fixed above the gear to hide it from view. At the bottom, the lower edge of the door is grooved to receive the metal channel which is secured by screws into the wood-work. The floor or bottom guide is of nylon and is screwed to the floor and in conjunction with the channel keeps the door in correct alignment. There should be a $\frac{3}{16}$ in. joint between the floor and the bottom of the door.

Fig. 2 shows the gear fixed directly to the lintel above the doorway, and a wooden pelmet is used in place of the metal one, which is supplied by the manufacturers as an optional extra. Fig. 3 shows the type of bracket required when the gear is to be fitted to the soffit of the opening. If double doors are required in this position, the brackets are fixed side by side as shown in Fig. 5. Fig. 4a is a view of the bottom nylon guide, which, is fixed near the jamb of the opening. Fig. 4b and c show alternative bottom guides to the door when these are favoured. Fig. 6 is a front view of the track and ball race.

Roller gear. Fig. 7 is a cross-section through the *Majestic* bottom roller gear. This is ideal for high class joinery items, and is most suitable for picture windows, Fig. 12, and doors to showrooms, etc. It is used for doors with a minimum thickness of $1\frac{7}{8}$ in. and a maximum weight of 600 lb. per door. The brackets holding the top box guide through which the rollers pass are fixed to the face of the masonry as shown, and if double doors are used, the brackets shown in Fig. 8 must be obtained. When double doors are to be placed beneath the soffit of the opening, the two box guides, Fig. 9, can be fixed to the woodwork covering the soffit.

The bottom rollers, which take the weight of the door, pass over a brass rail which can be recessed into a hard-wood or concrete sill, see Fig. 7. An alternative to this arrangement is seen in Fig. 10, where a brass rail, also used as a weather bar, has been let into the top surface of the weathered sill. To prevent draughts through the sides of the doors, sponge rubber buffers can be used, as in Fig. 11.

Small doors. For cupboard, cabinet and wardrobe doors up to a weight of 50 lb. per door *Loretto* cabinet rollers can be used, Figs. 14 and 16. These are so made that condensation will not affect them, and they are smooth and silent in their action. Fig. 14a, b and c show top guides with retractable shoots or bolts which enable the doors to be removed with the minimum of trouble. Fig. 16 gives two types of nylon top guides which are not retractable. That on the left is for edge fixing and that on the right is fixed from the top.

For frameless glass doors to cabinets, showcases, etc. an economical method of sliding the doors is by means of the track, Fig. 15d, and slider, Fig. 15a.

Garage doors. For two-leaf garage doors up to a maximum weight of 260 lb. per door, and where the doors are required to slide to each side of the opening as in Fig. 19, the straight sliding gear seen in Figs. 17 and 18 can be used. This consists of fixing brackets, which hold the track, to the inside or lower surface of the lintel above the opening. Two hangers are fixed to each door, and these pass along the track, as seen in the drawings. The channels in which the bottom guides travel should be let in flush with the concrete floor as this is being constructed. Two kinds of guides are illustrated, the roller type, Fig. 17, and that shown in Fig. 18.

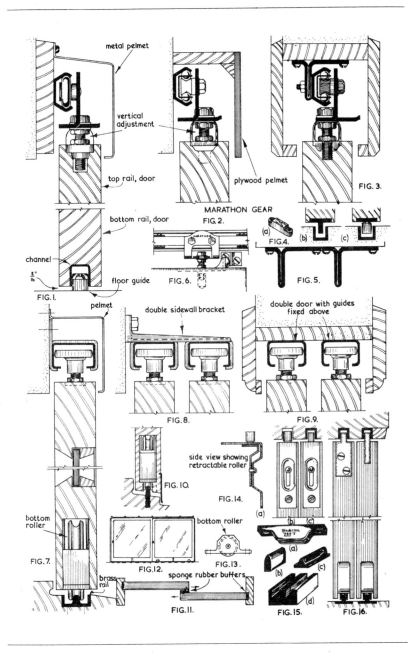

metal pelmet

vertical adjustment

top rail, door

plywood pelmet

FIG. 3.

MARATHON GEAR

FIG. 2.

bottom rail, door

channel

(a) FIG. 4. (b) (c)

FIG. 5.

floor guide

FIG. 6.

FIG. 1.

pelmet

double sidewall bracket

double door with guides fixed above

FIG. 8.

FIG. 9.

side view showing retractable roller

FIG. 10.

FIG. 14.

(a)

(b) (c)

bottom roller

bottom roller

FIG. 7.

FIG. 12.

FIG. 13.

(a)

brass rail

sponge rubber buffers

(b) (c)

(d)

FIG. 11.

FIG. 15.

FIG. 16.

251

The *Tangent* lock-up gear illustrated in Fig. 21 is suitable for garage doors of up to four leaves, a maximum height of 7 ft. 6 in., and a maximum weight of 70 lb. per leaf. Fig. 20 is a typical elevation of such doors. The section of the door shown in Fig. 21 is of a ledged and braced door for which the *Tangent* gear is most suitable. The track is fixed to the woodwork above the doorway opening by means of brackets, as shown in Fig. 22. A coupler bracket is used at the joint between two sections of the track.

Fig. 23 is a plan of the arrangement, and shows that the doors come to rest along a side wall of the building when in the open position. The hangers are fixed to the top rails or top battens of the doors and run along the track above the opening, Fig. 22. Hinges are fixed between the doors at the centre rail level, see Fig. 21, and the bottom guide and the channels at floor level.

The first leaf opens in the same way as a hinged door, see Fig. 23, this leaf being called the leading swinger. Bow handles should be fixed at three points along the centre rails so that these can be gripped to apply the pressure to slide the doors into the open or closed positions. As indicated in Fig. 21, the centre lines of the hangers, hinges and roller guides should all be kept in line.

Top-hung folding partitions. For these, Fig. 25, the *Council* end-folding gear is suitable. Fig. 24 shows the position of the gear where the doors are not hung below the soffit of the opening. The brackets holding the track are fixed to the lintel over the opening by means of bolts, and it will be seen that the knuckles of the hinges on the hanger side of the doors near the centres have to be kept out so that their centre lines are in line with the knuckles of the hangers and the guides. The hinges on the other face of the doors can be fixed in the normal fashion.

If the doors are to be kept flush with one face of the opening, the arrangement seen in Fig. 27 can be adopted. This involves having a rebated frame in the opening. Figs. 28 and 30 show how the edges of the leaves can be treated so as not to interfere with the hinges.

When the end leaf is a swinger, similar to a normally-hinged door, all the leaves of the partition should be equal in width, Fig. 29a, but where the end leaf carries a hanger it must be ¾ in. wider than the others, Fig. 29b. If the partition is to be directly under the soffit of the opening the brackets holding the track can be fixed as in Fig. 26.

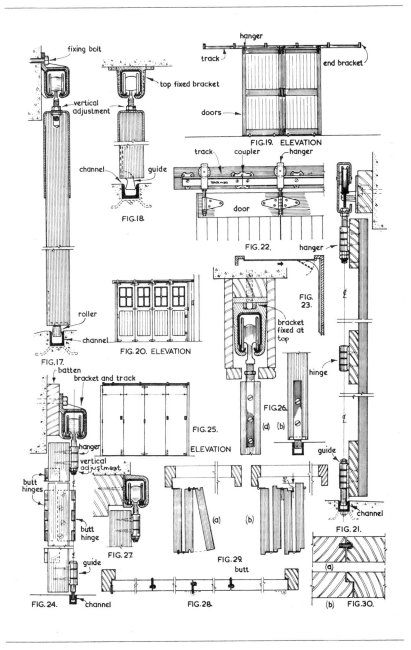

fixing bolt

hanger

track

end bracket

FIG.19. ELEVATION

doors

top fixed bracket

vertical adjustment

channel

guide

FIG.18.

track coupler hanger

door

FIG. 22.

hanger

FIG. 23.

bracket fixed at top

roller

channel

FIG.17.

FIG.20. ELEVATION

hanger

hinge

FIG.26.

(a) (b)

batten

bracket and track

FIG.25.

ELEVATION

guide

channel

FIG. 21.

hanger

vertical adjustment

butt hinges

butt hinge

FIG.27.

(a) (b)

(a)

(b)

FIG.29.

butt

guide

channel

FIG.24.

FIG.28.

FIG.30.

27 Windows with Curved Heads

Also see chapter 14.

Figs. 1, 2 and 3 show details of a semi-circular-headed frame with vertical sliding sashes. On the left of Fig. 1 the outside linings have been removed to show how the pulley stile has been fixed to the built-up head. The latter, which is semicircular, has been built up in three thicknesses, see Fig. 2, the centre section being the continuation of the parting bead, which is approximately $\frac{3}{8}$ in. thick and projects $\frac{5}{8}$ in. beyond the surfaces of the other two sections. The various segments of the head are glued and screwed together with the joints staggered.

The pulley stile is allowed to run up to well above the springing line of the frame, and is glued and screwed to the prepared flat surface of the head. A block shaped to fit between the stile and the head can, if necessary, be glued and pinned in position, as shown in Fig. 1. The two outer sections of the head are stopped about 4 in. above the springing line as the ends are required to act as stops for the sash. A small horizontal surface to correspond to the stop is provided, as in Fig. 7.

Fig. 2 is a section through the top portion of the frame and sashes, and it can be seen that blocks have been cut to fit between the linings above the built-up head to strengthen the work at this level. Fig. 3 is a horizontal section through the frame and a sash. Details around the sill are as shown in the companion volume, *Practical Carpentry and Joinery*. If it is required to have sliding sashes which act as pivot-hung sashes in addition to sliding vertically, double stiles must be provided for the sashes with stops between. These are then prepared as for the pivot-hung sash stops described in the companion volume.

Semicircular-headed frame with transom. Fig. 5 is a pictorial view of the top portion of a semicircular-headed frame involving a

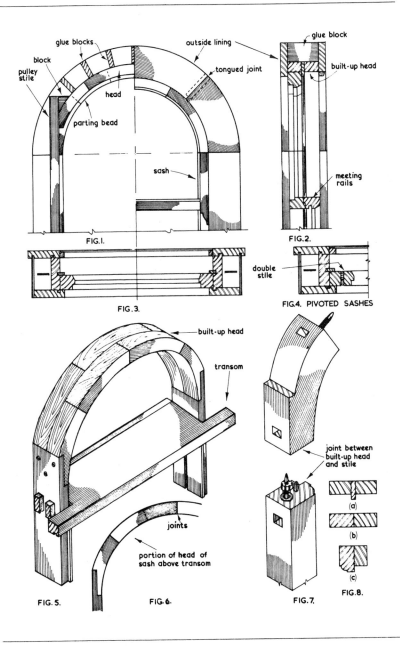

FIG.I.

glue blocks
block
pulley stile
head
parting bead
outside lining
tongued joint

FIG.2.

glue block
built-up head
meeting rails

FIG.3.

sash

FIG.4. PIVOTED SASHES

double stile

FIG.5.

built-up head
transom

FIG.6.

joints
portion of head of sash above transom

FIG.7.

joint between built-up head and stile

FIG.8.

(a)
(b)
(c)

transom and with sliding sashes below. The head of the frame in this case is built up of two sections only, a parting bead not being required above the transom. The latter is weathered on its top surface, and is kept down below the springing line so as not to complicate the joints around that section of the frame. Fig. 6 shows how the top part of the sash above the transom would be made.

Fig. 8 shows how the various shapes of heads of curved frames and sashes can be built up. At the top, a, are the three sections making up the head of the window in Fig. 1; b shows how the head of the frame in Fig. 5 is built up; and c indicates how the rebate can be formed in the semicircular portion of the sash shown in Fig. 6.

28 Roof Lights and Ventilators

Roof lights can be in the form of skylights, dormer windows, eyebrow windows and lantern lights.

Skylight. This consists of an opening in a roof, usually small, with a curb inserted around the opening. On top of the curb is placed the glazed sash which allows light to pass into the space below, see Fig. 1. The opening or trimming in the roof is made in the same way as an opening for a chimney, two of the common rafters and the trimming pieces forming the rectangular opening into which the curb is fitted.

Curb. The curb should be prepared from, say, $1\frac{1}{4}$ in. or $1\frac{1}{2}$ in. material, and can be either butt-jointed at the corners, or, if it has to be moulded on the lower inside edges, mitred or scribed at these points. The curb should be allowed to extend upwards so that its top edges are at least 3 in. above the surfaces of the roof covering, enabling it to be weathered in the correct manner, as seen in Figs. 1 and 2. The light is made so that it projects well beyond the four edges of the curb.

Condensation will inevitably settle on the lower surfaces of the glass panes at one time or another, and provision has to be made to allow this moisture to escape on to the roof surface. The top surface of the bottom rail of the light is recessed so that the panes of glass pass over and rest on only a portion of the top surface. Condensation grooves or recesses are made for each pane, approximately $\frac{1}{8}$ in. deep, allowing the moisture to escape between the glass and the bottom rail, Fig. 3.

The light can be made to open, which means that it will be hinged at its highest point, or it can be fixed. In the latter case it can be screwed down on to the curb from the top and the screw holes either pelleted or filled with putty. A close look at the drawings shows the additional work involved to ensure a watertight job. Figs. 3–5 show how the skylight is constructed.

Dormer window. If a living room is to be provided in a roof space it is also necessary to provide adequate light for the room in the shape of a dormer window or an eyebrow window. In either case it must be remembered that floor joists must be inserted in place of the usual ceiling joists, see Fig. 7. The depth of the joists will depend on the span.

Allowance for window frame. When constructing a roof which is to include a room, the roof rafters must be supported in a different way from the usual method. This involves constructing trussed or boxed purlins, see Fig. 16, Chapter 6, which will enable the rafters to be supported vertically.

As the roof is being constructed, the frame on which the window is to be supported should be made and placed in position so that the short rafters from below the window down to the eaves can be fixed. The height of this frame should be such that the sill of the window frame will be at least 6 in. above the top surfaces of the rafters. A $\frac{3}{4}$ in. thick apron piece is fixed to the frame and should extend down to the rafters. A 4 in. by 3 in. trimming piece will enable the common rafters above the opening to be fixed and extend up to the ridge.

The next step is to construct the roof, including the window frame, as the work proceeds. The window frame supports the 3 in. by 2 in. plates which, in turn, support the roof timbers. The plates are allowed to project beyond the window frame so as to obtain an overhang in front of the window. When the roof has been covered, the flooring in the roof space can be fixed to enable the studding forming the sides and cheeks of the dormer to be fixed.

Fig. 8 is a cross-section through one of the dormer cheeks, and shows that the 2 in. by 2 in. studs forming the side cheeks have been kept back flush with the inside edge of the roof plate, enabling 1 in. weatherboarding to be nailed below the plate on the outside. An eaves closing board is similar to a fascia, and is cut to fit round the roof rafters, so closing the opening between rafters at the eaves. Fig. 9 gives details around one of the jambs of the window frame. The frame material is from 4 in. by 3 in. The weatherboarding is allowed to overlap the ends of the window frame, and is finished with a cover fillet and quadrant.

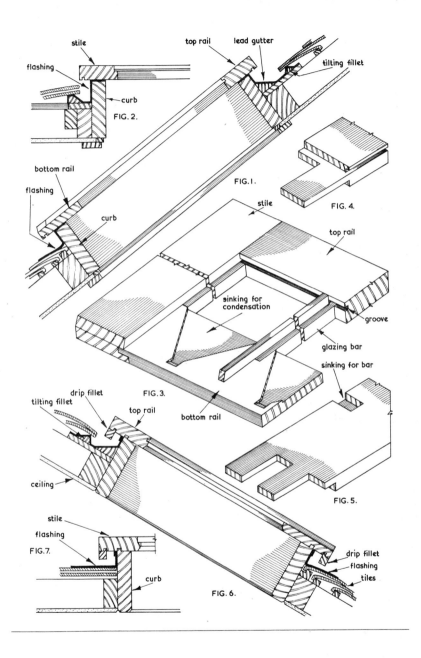

stile

flashing

top rail   lead gutter

tilting fillet

curb

FIG. 2.

FIG. 1.

FIG. 4.

bottom rail

flashing

curb

stile

top rail

groove

sinking for
condensation

glazing bar

sinking for bar

FIG. 3.

bottom rail

drip fillet

tilting fillet

top rail

ceiling

FIG. 5.

stile

flashing

FIG. 7.

curb

drip fillet

flashing

tiles

FIG. 6.

259

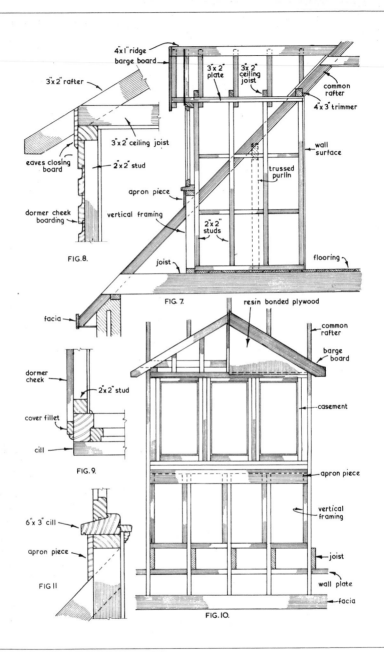

4"x1" ridge
barge board

3"x 2" rafter

3"x 2" plate
3"x 2" ceiling joist

common rafter

4"x 3" trimmer

3"x 2" ceiling joist

eaves closing board

2"x 2" stud

wall surface

apron piece

trussed purlin

vertical framing

dormer cheek boarding

2"x 2" studs

FIG.8.

joist

flooring

FIG. 7.

facia

resin bonded plywood

common rafter

barge board

dormer cheek

2"x 2" stud

cover fillet

cill

casement

FIG. 9.

apron piece

vertical framing

6"x 3" cill

joist

apron piece

wall plate

FIG II

facia

FIG.10.

Fig. 10 shows an elevation of the woodwork. Short 2 in. by 2 in. studs are fixed between the window frame head and the roof slopes to give a fixing for $\frac{1}{4}$ in. thick resin-bonded plywood which closes the space above the window. Barge boards fixed to the end rafters will give a finish to the appearance. Fig. 11 shows the details around the apron piece and the sill of the window frame. Before finishing the inside surfaces, the spaces between the studs and rafters should be well insulated.

Eyebrow window. These have become popular in recent years, but lack of knowledge has prevented many people from including this attractive feature in their own houses. First it must be realized that the steeper the pitch of the main roof rafters, the more suitable the roof is for incorporating an eyebrow window. A roof should be pitched at an angle of 60° or more to be suitable for the inclusion of a window of this type, though they have been constructed in roofs of much lower pitch.

Assuming that an eyebrow window has to be made in a roof pitched at $52\frac{1}{2}°$, Fig. 12, the first thing is to decide on the shape of the eyebrow profile, Fig. 13. This profile, which gives the shape of the eyebrow at the front, is built up in two thicknesses in a similar manner to the shaped rib of an arch centre. This can be made up in two halves and joined together on site. The shape of this profile is most

A view of an eyebrow window.

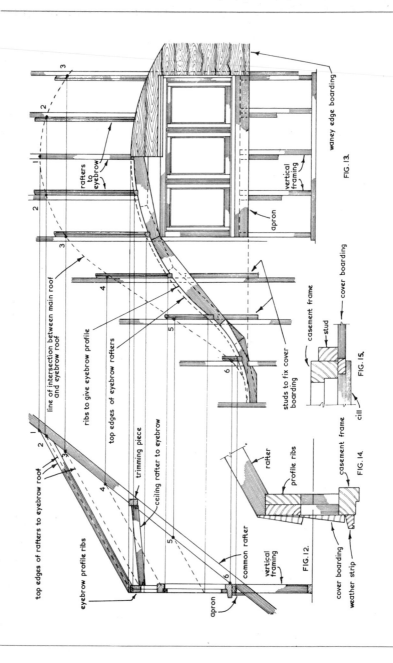

top edges of rafters to eyebrow roof

line of intersection between main roof and eyebrow roof

ribs to give eyebrow profile

top edges of eyebrow rafters

rafters to eyebrow

trimming piece

ceiling rafter to eyebrow

eyebrow profile ribs

studs to fix cover boarding

vertical framing

apron

waney edge boarding

FIG. 13.

casement frame

stud

cover boarding

FIG. 15.

cill

rafter

profile ribs

casement frame

cover boarding

weather strip

common rafter

vertical framing

apron

FIG. 12.

FIG. 14.

important, because the success of the work relies on its appearance. One should aim at a low, sweeping curve, rather than a high and rounded shape.

A vertical framing to carry the window frame, similar to that for the dormer window, should be prepared and fixed as in Figs. 12 and 13. Assuming that the opening in the roof has been completed, the profile should be offered up and fixed temporarily central to the position of the window frame, which, of course, must be made to fit between the vertical framing and the profile. Having positioned and fixed the window frame and the profile, prepare and fix the rafters to the eyebrow window. These rafters must all run from the profile board, backwards to the main roof, and must all be pitched at a pre-determined angle, say 30°. The eyebrow rafters in the drawings have all been notched over the top edge of the profile, and their edges are all, say, 1 to 2 in. above the top edge of the profile, see Fig. 13.

They are also fixed so that they can be secured to a common rafter at their top ends, see Fig. 13. If the eyebrow rafters are all set at the same angle the shape shown by the broken line in Fig. 13 will be automatically obtained. The broken lines in Fig. 12 show the lengths and positions of the eyebrow rafters. It will also be seen in Fig. 13 that a corner has been removed from the top edges of the rafters to give a better seating to the tile battens which flow over the eyebrow roof.

Figs. 12 and 13 give an indication of the geometry involved in setting out the eyebrow, and this should not be too difficult to understand. Vertical studs can be fixed on each side of the window frame to give support to the plywood, tiles, or waney edge-boarding to finish the front of the window. Fig. 14 shows details above the window frame and Fig. 15 the details at the side of the frame.

There is difficulty in making an eyebrow roof completely waterproof because a lot of water which settles on the surface will flow in a sideways direction. Extra efforts must be made, therefore, to stop the moisture which flows between the joints in the tiles getting to the roof timbers. Bitumen felt and/or building paper is often used for this purpose, with large overlapping joints, but it must be emphasized that this is the weakness of the eyebrow window. Finally, one last rule; keep the rafters to the eyebrow roof pitched as steeply as possible.

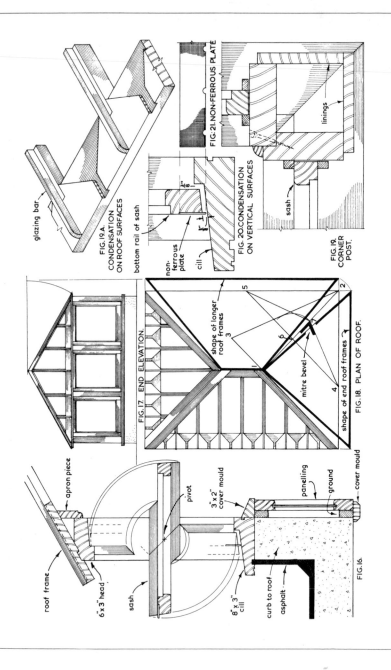

glazing bar

FIG.19.A. CONDENSATION ON ROOF SURFACES

FIG:21.NON-FERROUS PLATE

linings

sash

bottom rail of sash

non-ferrous plate

cill

FIG. 20-CONDENSATION ON VERTICAL SURFACES

FIG.19. CORNER POST.

shape of longer roof frames

mitre bevel

shape of end roof frames

FIG.18. PLAN OF ROOF.

FIG.17. END ELEVATION.

apron piece

roof frame

6"x 3" head

sash

pivot

3"x 2" cover mould

panelling

ground

cover mould

8"x 3" cill

curb to roof

asphalt

FIG.16.

Lantern lights. These, Fig. 16, are constructed to give light and ventilation to a space below a flat roof. The roof, usually of reinforced concrete, has an upstanding curb around an opening in the roof. The lantern light rests on this. Fig. 17 shows an end view of a lantern light. The vertical section through the light shows that the sill sits on the curb and overhangs at least $1\frac{1}{2}$ in. so that a drip can be worked near the front of the lower edge, and asphalt worked into a groove behind the drip to ensure a fully waterproofed job.

Four frames are prepared to form the sides of the light. One method of joining them at their corners is shown in Fig. 19. The ends of the sills are mitred and secured by means of one or two $\frac{3}{8}$ in. handrail bolts with one or two $\frac{3}{8}$ in. dowels. The jambs of the frames can be recessed into the sills and rebated together as shown. Screws and glue are used for fixing the jambs together down their length, and screws and glue again for fixing the jambs to the sill, the former passing through from the back edges of the jambs at an angle of about 60°. The linings forming the outside corner of the posts can be fixed to the jambs by means of glue and pins.

The roof of the lantern light, if it is hipped as in the drawings, is made in the shape of four frames mitred together at their intersections. A method of finding the shapes of the frames is given in Figs. 17 and 18 as is the method for constructing the mitre bevels. Those who do not follow these diagrams should refer to the chapter on roofing.

Turrets. These structures, Fig. 22, are usually situated on roofs of large buildings. Their function is to give a certain amount of ventilation to the building. Some are simple in construction, others are more elaborate and form an architectural feature as well as providing ventilation. Basically, a turret consists of four posts which pass through the surfaces of the roof down to some means of anchorage. In the example shown in Fig. 22 the posts are secured to beams which run between two of the trusses supporting the roof. Above the roof surfaces, and up to the sills, the posts have infillings of studs and nogging pieces, and can be covered internally and externally with resin-bonded plywood. The outside surfaces would later be covered with lead, copper or some other sheet metal.

Above the plywood surfaces are four frames with louvre boards, and these frames have been recessed into the sides of the four main

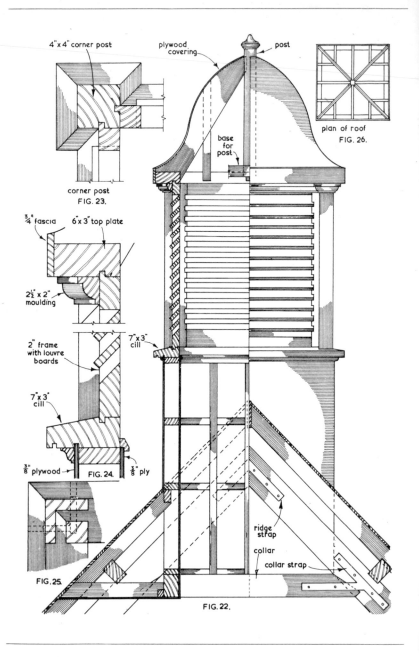

4" x 4" corner post

plywood covering

post

plan of roof
FIG. 26.

corner post
FIG. 23.

¾" fascia

6" x 3" top plate

base for post

2½" x 2" moulding

2" frame with louvre boards

7" x 3" cill

7" x 3" cill

⅜" plywood

FIG. 24.

⅝" ply

FIG. 25.

ridge strap

collar

collar strap

FIG. 22.

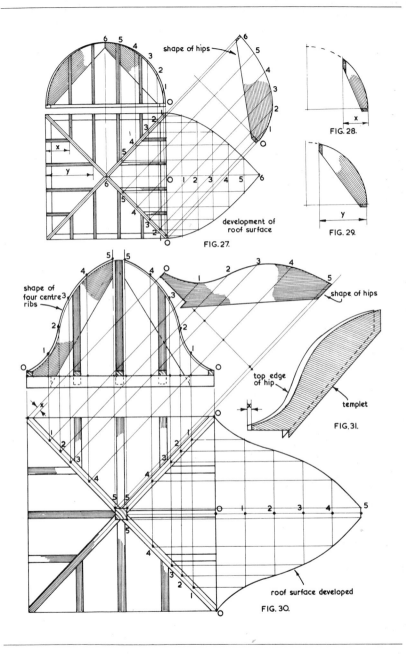

shape of hips

x
y

O 1 2 3 4 5 6

development of
roof surface

FIG. 27.

FIG. 28.

x

FIG. 29.

y

shape of
four centre
ribs

shape of hips

top edge
of hip

templet

FIG. 31.

x

O 1 2 3 4 5

roof surface developed

FIG. 30.

posts. On the left of Fig. 22 is a section through the turret, and on the right an elevation.

Fig. 26 is a plan of the turret roof and is simple in design. The centre post of the roof is secured at the bottom to a base plate which extends across to two of the top plates. The centre post passes through the centre of the turret roof, continues some distance above the surface, and is shaped to give a finish. Figs. 23, 24 and 25 give larger details of the construction of the turret. Fig. 23 shows how the louvred frames are tongued and grooved to a corner post, Fig. 24 is a section through one of the frames and Fig. 25 shows a corner of the sill mitred and dowelled to a corner post.

Before leaving the topic of roofs with shaped ribs, two examples are given in Figs. 27 and 30. They indicate methods for developing the shapes of the various parts. Fig. 27 is the plan and section through a roof which is semicircular in elevation and square in plan. The shapes of the ribs can be seen in the elevation but the shape of the hips must be developed.

Divide one half of the elevation into a number of parts, say six, and project these points down to the plan and on to the centre line of one of the hips. From these points draw lines at right angles to the hip, and from a base line make the various lines brought up from the hip equal in length to those in the elevation. A curve drawn through the points obtained will give the shape of the top edges of the hips.

To obtain the shape of the plywood required to cover one surface, draw lines out from the points on the two hips in the plan, as shown, and on the centre line mark off distances equal to those round the half of the elevation. Draw vertical lines through these points to intersect with those brought over from the hips in the plan to give points on the outline of the plywood shape.

Figs. 28 and 29 show the shapes of the short ribs seen in the plan. In each case the radius used is that used for the elevation.

Ogee shape. Fig. 30 is the plan and elevation of an ogee roof, similar to that used for the turret, Fig. 22. The geometry for this roof is very like that of the semicircular roof, Fig. 27. The only problem is how to mark the backing on the top edge of the hips, Fig. 31. A templet is made exactly the shape of the hips. It is placed on each hip in turn and allowed to slide sideways the distance x in the plan, Fig. 30. The outline of the templet will give the outline of the backing.

29 Panelling to Walls

Although wall panelling is fast disappearing in domestic buildings a great deal is still being carried out in public buildings such as magistrates courts, town halls, director's suites in large commercial buildings, and so on. Most of the panelling today extends up to frieze height, which is anything above the height of the doorways, or to the full height of the walls up to the cornice. Years ago most of this work was dado panelling, which extended up to the height of the tops of chairs so that these would not damage the plaster work when pushed up against the walls.

Dado panelling. A vertical section through and an elevation of some dado panelling is shown in Fig. 1 and is approximately 3 ft. 6 in. high. It consists of panelled frames to any required design secured to wood grounds carefully fixed to the wall in vertical and horizontal alignment. The grounds are usually fixed to wood plugs which have been inserted in vertical joints of the brickwork, the walls above the top grounds being plastered before the panelling is fixed.

Fig. 2 shows, to a larger scale, a vertical section through the dado panelling, and it will be seen that the wood plugs have been inserted into the brickwork joints and their front edges trimmed so that they are in perfect alignment. The grounds are fixed to the plugs, and the wall above the panelling is plastered. The panelling is fixed by screwing into the grounds, and a capping on the top and a skirting at floor level completes the work. If a moulded strip is first screwed to the floor and the tongue on the bottom edge of the skirting allowed to fit into a groove on the moulded strip, the skirting need be fixed at the top only, so allowing movement to take place in the timber without creating an open joint between the bottom of the skirting and the floorboards.

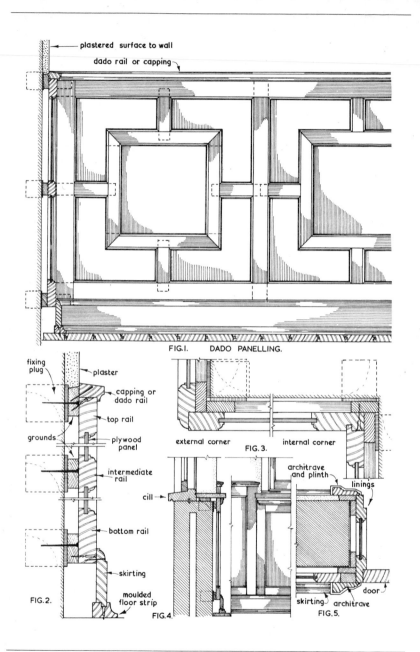

plastered surface to wall

dado rail or capping

FIG.1.   DADO PANELLING.

fixing plug

plaster

capping or dado rail

top rail

grounds

plywood panel

intermediate rail

bottom rail

skirting

moulded floor strip

FIG. 2.

external corner   FIG. 3.   internal corner

cill

FIG.4.

architrave and plinth

linings

skirting   architrave

door

FIG.5.

Fig. 3 shows how the edges of the panelled frames can be treated at their junctions at internal and external corners around the room. Fig. 4 indicates how the framing can be fixed when involving a sill board at a window opening, and Fig. 5 gives details of dado panelling where it intersects with the framed door linings.

Frieze height panelling. Fig. 6 is a vertical section through wall panelling which extends up beyond the dado to the frieze rail. The panelling up to the dado height can, if required, be treated as a separate unit with the dado rails covering the gaps between the lower and upper panelled frames. The frieze rail can be treated in several ways, two being shown in Figs. 6 and 6a. In Fig. 6 are seen details where concealed lighting is to be a feature in the room. A trough, reinforced with purpose-made mild steel brackets, is constructed and in this is placed the wiring and the fluorescent tubes or bulbs. In Fig. 6a a small shelf is provided on which is placed such things as brassware, etc.

Clearing projections. Often there are projections such as piers in long walls, and Fig. 7 shows how these are negotiated. A pilaster is prepared to cover the face of the projection, and small plain side pieces fit in between the pilaster ends and the main panelling.

Where long lengths of panelling are required the frames are built up in sections and the joints between the frames can be covered with a pilaster as in Fig. 8. This, as will be seen in the drawing, is fixed straight on to the faces of the main panelled frames. Fig. 9 shows methods of finishing the external corners of pilasters and Fig. 10 is a pictorial view of the pilaster in Fig. 8.

Methods of fixing. There are several methods of fixing panelling, including pilasters. Often the fixing is carried out by screwing through the face of the framework and filling the holes with pellets made from a similar timber, Fig. 11a. Figs. 11, 12 and 13 illustrate other ways of fixing panelling to the grounds used in better class work. The finished work will show no signs of the fixings; neither will it be obvious at which points the fixings have been used.

In Fig. 11 a rebate has been worked on the top back edge of the horizontal grounds, and lipped buttons are screwed to the back surfaces of the panelled frames. The frame is fixed by 'hanging', which involves the lips of the buttons engaging the rebates in the grounds.

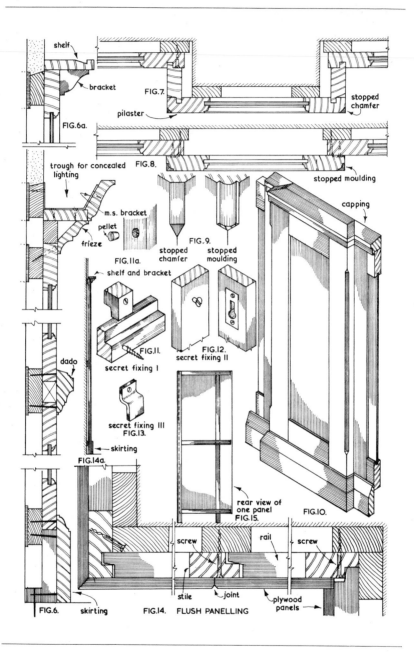

shelf

bracket

FIG.7.

pilaster

stopped chamfer

FIG.6a.

FIG.8.

stopped moulding

trough for concealed lighting

m.s. bracket

pellet

frieze

FIG.11a.

stopped chamfer

stopped moulding

FIG.9.

capping

shelf and bracket

FIG.11.

secret fixing I

FIG.12.

secret fixing II

dado

secret fixing III
FIG.13.

rear view of one panel
FIG.15.

skirting

FIG.14a.

FIG.10.

screw

rail

screw

FIG.6.   skirting

stile

joint

plywood panels

FIG.14.   FLUSH PANELLING

When in position the frames are secured to prevent them from becoming dislodged from the grounds. This is done by inserting screws through the frames into the grounds at points where the screws will not be seen, for instance, behind where the skirting will be fixed, and at the top where the screws will be hidden by the capping or frieze rail. Instead of wood buttons being used the mild steel button, Fig. 13, can be used. Fig. 12 shows another method, based on the slot and screw. Screws, with their heads projecting are fixed to the grounds at convenient points and plates, similar to that in Fig. 12 are secured to the back surfaces of the panelling, to coincide with the positions of the screws in the grounds. When the panelling is fixed, the heads of the screws are allowed to enter the holes in the plates and the panelling is then carefully tapped downwards, so that the screw shanks enter the narrower slots in the plates.

Figs. 14 and 15 show a method of fixing flush plywood panels. At Fig. 15 is shown the back surface of one of the panels. It has had a frame glued to the surface, each frame consisting of one stile and three rails. The stile overlaps the edge of the panel by, say, 1 in., and is rebated on the edge nearest the grounds, see Fig. 14. The ends of the rails farthest from the stile are cut to form a tongue which will fit into the rebate on the stile of the next panel, Fig. 14. By working round the room in the appropriate direction each panel can be fixed by first inserting the three tongues in the rebate of the previously fixed panel and then fixing the other edge of the panel by screwing through its stile into the grounds. Also on the drawing, Fig. 14, is shown how external and internal corners can be treated.

30 Counter Construction

The first of two methods for constructing counters are shown in Figs. 1 and 2, which are a vertical cross-section and part elevation of what could be described as a traditional style counter. The three components are a top of solid timber, a framed and panelled front with raised and fielded panels and bolection mouldings, and the pedestal or cupboard section. This counter is suitable for a bank or showroom.

The top, of solid hardwood timber such as mahogany, is $1\frac{1}{4}$ in. thick, and this has been increased in the front by another piece $1\frac{1}{4}$ in. thick to give the appearance of its having been prepared from much heavier timber, see Fig. 5. The panelled front is made up in sections (depending on the overall length of the counter) with shaped, solid pilasters covering the joints between the panelled sections. The cupboard or pedestal sections can be made up as separate units.

Fig. 6 is a pictorial view of a pedestal and comprises drawer and cupboard spaces. These units can be made from softwood and plywood, and painted or stained to match the hardwood top and front. The spaces between the units can be utilised as necessary. If clerks are to sit behind the counter these are adapted as knee spaces.

Modern style counter. Fig. 4 is a vertical section through a more modern type of counter, the top and front being made almost entirely from blockboard. The joints between the various panels which form the front of the counter can be masked by plain pilasters and the top built up at its front edge as in Fig. 7. The pedestal, which is also made from blockboard, sits on a plinth frame made from $1\frac{1}{2}$ in. timber strengthened by glue blocks.

Figs. 8 and 9 are cross-sections through the pilasters in Figs. 2 and 4. The top and front of the blockboard counter are often surfaced with a plastic such as *Formica,* and if the surfaces of the pilasters are kept flat, these, as well as the skirting, can also be covered with the same or a contrasting plastic.

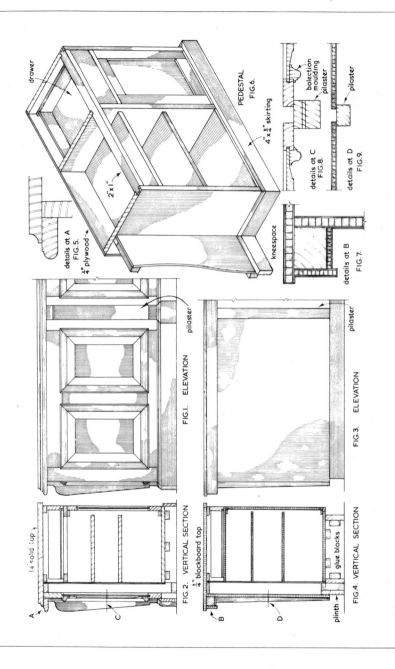

drawer

2"x1"

details at A
FIG.5.
¼" plywood

PEDESTAL
FIG.6.

4"x ½" skirting

kneespace

bolection
moulding
pilaster

details at C
FIG.8.

pilaster

details at D
FIG.9.

details at B
FIG.7.

pilaster

FIG.1.   ELEVATION

pilaster

FIG.3.   ELEVATION

1¼ solid top

FIG.2.   VERTICAL SECTION
¾" blockboard top

A

C

glue blocks

FIG.4.   VERTICAL SECTION

B

D

plinth

31 The Construction of Stairs and Handrails

Also see chapter 17.

Straight flights, dog-legged stairs, and open-newel stairs have already been dealt with in the companion book, *Practical Carpentry and Joinery*. The types to be described in this chapter are known as geometrical stairs. These do not contain newel posts and therefore have to be constructed with continuous strings and handrails.

Details of a geometrical staircase. Let us first consider the general layout of a geometrical stair. Fig. 1 is the plan of such a stair, and comprises two straight flights of five steps in each, these flights being connected by three winders where the stairs change direction. The first step in a traditional geometrical stair is usually shaped like the one shown on the plan which is called a curtail step. The shaped end generally follows the curve of the handrail scroll which is immediately above it.

A variation to the curtail step is the commode step, the outline of which can be seen in broken line. This is constructed in the same way as the curtail step but it also has a curved riser which goes across the full width of the stairs. The plan of the stairs is only one of many layouts that it is possible to get with geometrical staircases. For instance, the two straight flights can be connected with a quarter-space landing instead of the winders; the turn of the stairs can be 180°, not 90° as shown; also the two parallel flights can be connected by a half-space landing; or a half-turn of winders, or even a quarter-space landing and a quarter-turn of winders.

String details. The wall strings are constructed similarly to those for dog-legged and open-newel stairs, but the outer strings are often cut to the outline of the steps, as seen in Figs. 2 and 3. These cut strings, as they are called, are prepared so that the treads actually rest on the top horizontal surface of the step outline, and the risers are

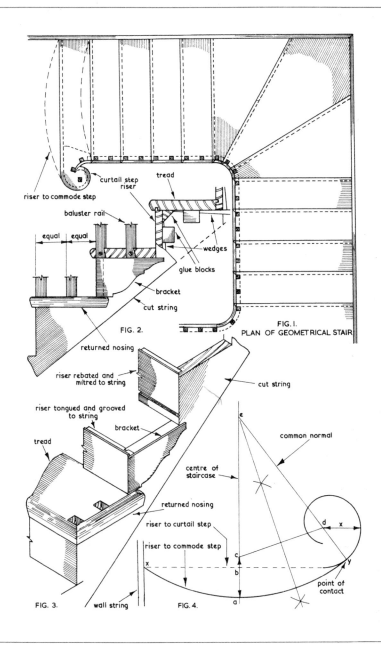

curtail step
riser

riser to commode step

baluster rail

equal | equal

tread

wedges

glue blocks

bracket

cut string

FIG. 2.

returned nosing

riser rebated and
mitred to string

cut string

riser tongued and grooved
to string

bracket

tread

returned nosing

riser to curtail step

riser to commode step

FIG. 3.

wall string

FIG. 4.

FIG. 1.
PLAN OF GEOMETRICAL STAIR

common normal

centre of
staircase

point of
contact

either mitred or tongued-and-grooved to the front vertical edge, see Fig. 3. If the risers are tongued-and-grooved to the string, the end grain of the riser is concealed with a shaped bracket. This is a thin piece of solid timber or plywood, glued and pinned to the surface of the string. No bracket is required if the riser is mitred to the string.

The top step in Fig. 2 shows details of how the treads and risers are tongued-and-grooved together. These details are slightly different from those for steps in closed or uncut strings, being more practical for the cut-string type of stair.

The baluster rails are dovetailed to the ends of the treads and glued and screwed into position, the joints being hidden by the return nosing which is mitred to the front nosing of each step, see Fig. 3. The method used for correct positioning of the baluster rails is shown in Fig. 2. As many glue blocks as necessary should be used for strengthening the steps.

Up to now we have not mentioned the portions of the strings which involve the change of direction, neither have we dealt with the shaped steps. Let us consider the curved portion of the strings first.

Curved strings. Fig. 11 deals with the turn in the staircase at the top of the first flight, but before we consider this problem we should first take a simpler example. Fig. 9 shows how these curved portions of the strings are constructed. We have to make a drum or former on which the timber can be bent to the required shape. These curved portions of the strings are built up to any required thickness with veneers from $\frac{1}{16}$ in. to $\frac{1}{8}$ in. thick. The first of the veneers to go on to the drum should have the marking-out lines transferred from the rod to its inside surface, see Fig. 8. The veneer is then carefully positioned, bent round the drum, and fixed in position with clamps or G cramps, Fig. 9. Other veneers are added to the first, say three or four at a time, the surface of every veneer being covered with a cold-water glue such as casein or some suitable synthetic resin glue. Each veneer must be kept in close contact with the surfaces of the adjacent veneers, additional clamps being used as necessary for this purpose. When the glue has hardened the built-up string can be removed from the drum and, by following the lines which were placed on the first veneer to be placed on the drum, it can be shaped and prepared as seen in Fig. 10.

Fig. 5 shows how a continuous string for a staircase of two flights

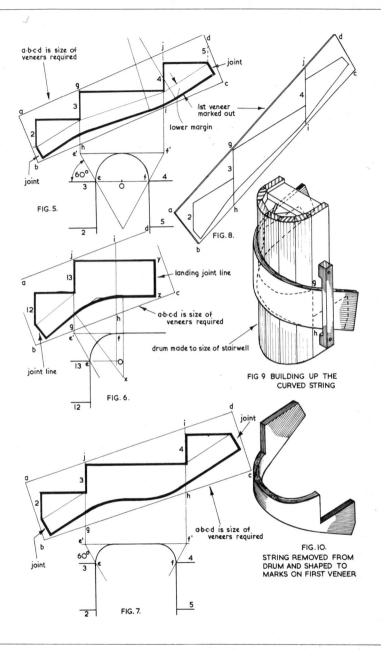

a·b·c·d is size of veneers required

joint

g

5

4

3

joint

2

b

joint

60°

e

O

f

4

3

2

d

5

FIG. 5.

FIG. 8.

d

j

c

4

i

g

3

h

a

2

b

1st veneer marked out

lower margin

j

i

13

y

landing joint line

a

12

c

g

h

a·b·c·d is size of veneers required

b

e'

f

joint line

13

e

O

12

x

FIG. 6.

drum made to size of stairwell

FIG 9 BUILDING UP THE CURVED STRING

d

joint

i

j

4

3

c

2

h

b

g

a·b·c·d is size of veneers required

joint

e'

f'

60°

e

f

4

3

5

2

FIG. 7.

FIG. 10.
STRING REMOVED FROM DRUM AND SHAPED TO MARKS ON FIRST VENEER

connected by a half-space landing should be set out. The lower part of the drawing shows the plan of the stairs. Risers 2 and 3 are the top steps of the lower flight, and risers 4 and 5 the lower risers of the top flight. The space between risers 3 and 4 is the half-space landing.

Preparing a stretch-out. To obtain what is called the stretch-out of the steps, two 60° lines should be drawn, one through each end of the diameter e–f to intersect the horizontal line just touching the top of the curve in e′ and f′. The distance e′–f′ is the horizontal distance from e round to f seen in the plan. Project e′ and f′ upwards vertically so that the stretch-out of the steps can be constructed as shown. The only dimensions required to draw the stretch-out are the 'going' and the 'rise' of each step.

Having drawn the outline of the steps in the stretch-out, straight lines connecting the lower ends of the risers should be drawn and the distance known as the lower margin marked parallel to these lines. This lower margin can be any dimension, and depends on the construction of the stairs. If, for instance, the stairs are fairly wide, say 4 ft. 6 in. to 5 ft., it may be necessary to support the centres of the steps with a carriage and a series of brackets, see Fig. 14a.

As it is necessary to add a soffit to the underneath surface of the staircase it is also essential to have the lower margin wide enough to come below the lower edge of the carriage pieces so that the soffit is supported.

Joints. It will be noticed that the joint at each end of the stretch-out is one step behind the end of the curved portion. For instance, at the lower end of the marking out it can be seen that the joint is immediately below riser number 2, but the string does not begin to turn until it gets to riser number 3. At the top end the same thing occurs; the curve of the string stops at riser number 4, but the joint is immediately below riser number 5. These joints are where the section of the curved string is joined to the straight strings of the lower and upper flights.

The type of joint used is the counter cramp, see Fig. 14. Before placing the veneer which has been set out round the drum, it must be ascertained that the guide lines g–h and i–j are on the veneer. These will assist in placing the veneer on the drum in the correct position. If they coincide with the ends of the curved portion of the drum and are in the vertical direction, it is fairly safe to assume that the 'rise' or position of the veneer is correct.

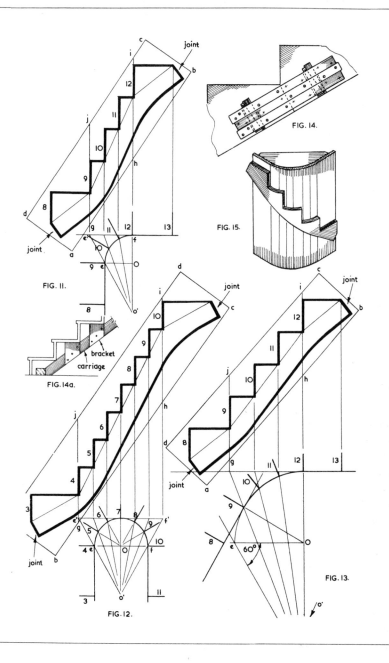

FIG. 11.

FIG. 14.

FIG. 15.

FIG. 14a.

FIG. 12.

FIG. 13.

281

Fig. 10 shows what the curved portion of string will look like when it has been removed from the drum and shaped.

Fig. 6 is a plan of a straight flight rising to a landing, riser number 13 being the last one on the stair. As the turn in this example is of 90°, a 60° line through point e to intersect the horizontal line in e′ will give the distance round the curve from e to f, the developed distance being e′–f. Remember that the curved portion of string this time is jointed to the front apron piece which forms the face of the landing timbers. Therefore from riser 13 on the development, the direction of the marking out will be horizontal, following the line of the landing. The distance between points y and z will be equal to the depth of the apron piece along the front of the landing.

Fig. 7 is an example similar to that in Fig. 5, two straight flights connected by a half-space landing. The distance e′–f′ is equal to the distance e–f on the plan. Remember, lines h–i and g–j must be placed on the first veneer in every case.

Quarter space of winders. We now come to a turn in a stair similar to that in the staircase in Fig. 1. Two straight flights are connected by a quarter-space of winders (see Fig. 11). The distance e–f is developed in the usual way by drawing a 60 degrees line through e, the dimension e′–f being the distance round the turn from e to f. As we also require the positions of risers 10 and 11 on the stretch-out, the 60° line should be extended downwards to meet the vertical line from o in o′. Lines from o′ should be drawn through the ends of risers 10 and 11 to give their positions on line e′–f. It should be fairly clear what should be done from this point to complete the marking out. Fig. 15 shows the completed portion of the string put back on the drum until it is required.

Half-space of winders. Fig. 12 shows the setting out for two flights connected by a half-space of winders, and the drawings should be clear enough to follow if the foregoing notes have been memorised. Fig. 13, however, shows two flights connected by a one-third space of winders, or in other words, two flights connected by winders giving the stairs a 60° turn.

The plan of the stairs should be constructed, and the curved portion can, temporarily, be changed into a 90° turn as shown. A 60° line from e downwards to meet a vertical line from o to give point o′, and lines from o′ through the ends of risers 9, 10 and 11 will give the positions of these risers in the stretch-out.

Shaped bottom step. We should now turn our attention to the shaped steps at the bottom of the staircase. Consider the curtailed step first, Fig. 20. It is necessary to know the shape of the scroll which is situated immediately above the shaped end of the step. The outline of this is first constructed (see Fig. 11 Chapter 19), and the setting out of the shaped end of the curtail step superimposed over the scroll lines, making the nosing line of the step follow the outside edge of the scroll. As can be seen from the drawings, a block is first made similar to the shape taken from the setting out (Figs. 20 and 22). This can be built up from several pieces of timber with the grain going in opposite directions like a piece of plywood. If a bandsaw is available the shape of the block is best cut on this.

A modern staircase with a centre laminated string. The treads are cantilevered on each side of the string and are fixed with purpose made brackets. The string is approximately 28 ft. long, 18 inches wide and 15 inches deep, the laminations are of 1 inch Douglas Fir and were glued together with 'casco' casein glue M1562.
Elliott, White, Reading.

283

The block is fixed to the end of the string by means of a tongued-and-grooved joint, and is glued and screwed. Recesses should be made in the block to receive a pair of folding wedges near the centre of the block. These are used for fixing the two veneers which meet at that point; and another recess is needed for receiving the riser less the thickness of the veneer at the front of the block.

The short veneer should first be fixed by gluing to the block, and this can be accomplished by the use of a former block and G cramps. The former block is cut to fit up against the veneer when it is in position, and the G cramps will hold the veneer tight up against the block until the glue has set.

The longer veneer is obtained by reducing one end of the first riser to between $\frac{1}{16}$ in. and $\frac{1}{8}$ in., taking care to keep the thickness uniform throughout its length. When the G cramps and block have been removed the second and longer veneer can, if necessary, be steamed or soaked in hot water to soften the fibres. The veneer and block surfaces are glued, and the end of the veneer entered into the recess for the folding wedges near the centre of the block. The wedges should then be glued and driven home, care being taken to see that the veneer is positioned correctly so that its edges will run parallel to the block when it is wrapped around into its final position.

To enable the surfaces of the block and veneer to come in to close contact with one another, folding wedges are glued and inserted between the shoulder at the end of the veneered portion and the block and carefully driven home. Screws can be used for keeping the remainder of the riser up against the block, as seen to the left of Fig. 20. The tread to the shaped step is prepared to the shape of the outline of the scroll. This is the nosing line indicated on the plan, Fig. 20.

Commode step. If this is to be constructed, the manner described for the curtail step can be followed with slight variations. The block requires two formers, one at the top and one at the bottom, recessed into it so as to form the curved riser which goes right across the width of the stairs, see Fig. 21. Saw kerfs are used in bending the thicker portion of the riser. Some difficulty, however, may be found in setting out the shape of the step, see Fig. 4. Let a–e be the centre line of the staircase, a–b the amount the curved riser is in front of its position if it were a curtail step, and a–c the radius of the last curve used when setting out the scroll. This is equal to d–y, Fig. 4. Join c and d and then

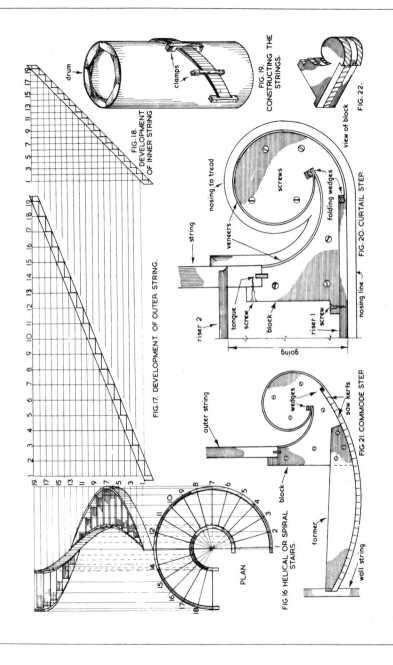

FIG. 18. DEVELOPMENT OF INNER STRING

drum

clamps

FIG. 19. CONSTRUCTING THE STRINGS.

view of block

FIG. 22.

FIG.17. DEVELOPMENT OF OUTER STRING.

nosing to tread

string

veneers

folding wedges

screws

riser 2

tongue

screw

block

riser 1

screw

going

nosing line

FIG. 20. CURTAIL STEP.

outer string

wedges

saw kerfs

block

former

wall string

FIG. 21. COMMODE STEP.

PLAN

FIG 16 HELICAL OR SPIRAL STAIRS.

285

bisect this line to give e on the centre line of the stair. Use centre e for drawing the curve y–a–x.

Spiral stairs. We now come to a different type of geometrical stair, namely, the spiral stair or, more correctly, the helical stair. In the plan, Fig. 16, the stair appears to be circular, and the elevation shows that it is similar to the thread of a screw or bolt. The strings to a helical staircase are constructed in the same way as for the type of staircase already dealt with. They are built up to any required thickness by using a number of veneers. In a staircase such as that in Fig. 16, the thickness of the veneers could be as much as $\frac{1}{4}$ in. A drum, similar to that shown in Fig. 19, has to be made to the required dimensions, and veneers placed round it and clamped in position until the glue has set. The first veneer to be placed on the drum in the case of the outer or longer string, and the last veneer in the case of the inner or smaller string, must have the marking-out lines placed on them, as for the continuous string of the other type of geometrical stair.

It should be realized that to set out the strings to a helical stair, it is not necessary to draw the elevation of the staircase as shown in Fig. 16. This drawing has been included to illustrate the type of stair we are dealing with. All that is needed is the plan of the stair and knowledge of the rise of each step.

String development. To develop the outside or larger string to the stair, first draw the plan and number the risers 1, 2, 3, etc. On a horizontal line mark off the distances 1–2, 2–3, 3–4, etc. These dimensions can be taken from the plan. Remember that it is the first veneer of this string which must have the marking out placed on it, so the distances 1–2, 2–3, etc. are taken from the inside edge or the concave edge of the string. Mark these distances 1–19 on the horizontal line, Fig. 17, and drop vertical lines from all these points to intersect with horizontal lines brought over from the vertical line, giving the heights of all the risers to the stair.

This vertical line is similar to a storey rod and can be seen to the left of Fig. 17. Draw in the outline of the steps obtained and then mark in the top and bottom edges of the string to any convenient size. As can be seen, the string is a straight piece of timber, and has only to be placed round the drum at the correct rake or angle to produce the correct curvature. In the workshop the setting out of the string would be done on the surface of the drum and not on a large flat surface because this would probably not be practicable. The foregoing description is included to enable readers to understand the geometry involved.

The inner string is made in the same way. Another drum, to the required dimensions, is made, and the veneers to form the small string are cramped round it as for the larger string. As it is the last veneer which must have the marking out placed on its surface, the distances 1–2–3 etc. on the horizontal line have to be taken from the plan of the small string, Fig. 16, and on the edge nearest the outside or large string. This veneer will be clamped on to the drum last of all, and straight on to the top of the others already in position. Remember to place the veneers round the drums at the correct rake. It is quite simple to work out how high the top edge of the strings have to rise in passing round the drum the required distance. For instance, the strings in Fig. 16 have to travel upwards a distance of 19 risers in passing round the drum three-quarters of a turn. Remember, also, that the top edges of all the veneers have to lie in a horizontal plane when tested radially.

The steps to a helical stair can be of the traditional kind or can be of the open-riser type, which means that treads only are to be used. The strings can be recessed to take the ends of the treads or treads and risers in the usual way. Some modern staircases constructed in the helical style have a central string and the treads fixed to the string by means of wooden or purpose-made metal brackets. Sometimes these strings are as much as 18 in. to 2 ft. in depth. They are built up with $\frac{1}{2}$ in. thick veneers, three or four being placed round the drum at a time, these being left for twenty-four hours when another four are added, and so on until the required thickness has been reached.

Handrailing. We now come to the problem of handrailing. As no newels are included in geometrical stairs, the handrails have to be continuous like the strings. To form the curved portion of handrail at the change of direction in the traditional way of handrailing, however, demands a greater knowledge of geometry, especially in handrailing involving two bevels.

Single-bevel work. The first two examples are single-bevel work and are not very difficult when compared with two-bevel work. When we consider all the other examples in this chapter, however, the reader will have to turn to Chapter 19 and learn how to develop oblique planes if he wishes to understand what is happening in this part of the work on handrailing. Let us consider the example in Fig. 23. At the bottom of the illustration is shown the plan of a staircase at the top of a straight flight of stairs leading to a landing. The handrail at this point is also shown. Above this drawing is the elevation of the last two steps in the flight which will, of course, give the pitch of the stairs. Notice that the curved portion of handrail in the plan is placed centrally over the quadrilateral a–b–c–d. This quadrilateral is the plan of a square prism with its top surface inclined at an angle, in this case equal to the pitch of the stairs (Fig. 26).

Preparing templets : To develop the shape of the templet required to make the curved part of the handrail, it is first necessary to develop the shape of the top surface of the prism, and from that develop the shape of the templet. Proceed as follows. Draw the plan of the handrail and the elevation of the top steps. Place in position the plan of the square prism so that sides b–c and c–d are in line with the centre lines of the handrail. Draw the pitch line of the stairs, and project up to the pitch line the edges a–d and b–c of the prism to give a and b on the pitch line. Project these points over at right angles and make a–d and b–c in the development equal to a–d and b–c in the plan. Then a–b–c–d in the development is the true shape of the top surface of the prism.

Divide the width of the plan of the prism into any number of parts and project these points up to the pitch line, and from here over to the c–d edge of the development. Make the distances a–a′, a–a″, 1–1′, 1–1″, etc. equal to those in the plan. Draw freehand curves through these points to obtain the shape of the templets required. Add two straight sections, say 2 in. long, on each end of the curved templet to complete its shape.

Two templets this shape are required for the wreath (this is what the curved portion of the handrail is called), and these can be cut from, say, $\frac{1}{4}$ in. plywood, see Fig. 25a. The thickness of the material required for the wreath is found by drawing a horizontal line to represent the top edge of the material (25c). Draw a line at any point at the same angle as the bevel shown in the elevation, Fig. 23, and from this angle construct a rectangle equal to the size of the handrail. Draw another horizontal line through the bottom edge of the handrail, and the distance between the two horizontal lines is the minimum thickness of timber required for the wreath.

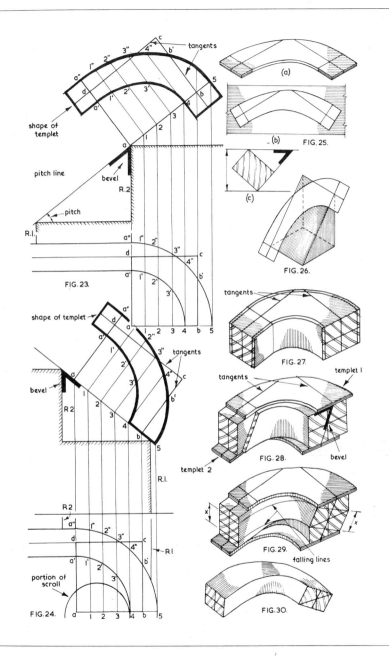

tangents

(a)

(b) FIG. 25.

(c)

FIG. 26.

c
4" b'
3"
2"
1"
a"
d
a'
a
5
3' 2' 1'
b
4
shape of
templet
3
2
pitch line
1
bevel
R.2
a
pitch
R.I.
FIG. 23.

a" 1" 2"
d 3"
4" c
a' 1' 2' b'
3'
1 2 3 4 b 5

tangents

FIG. 27.

tangents   templet I

FIG. 28.   bevel

templet 2

shape of templet
a"
d
a
a'
1'
2"
1"
tangents
2'
3"
3'
4"
4
c
b'
b
5
bevel
a
R 2
1
2
3

R.I.

R 2
a"
1"
2"
d
3"
4"
c
a'
1'
2'
b'
3'
portion of
scroll
FIG. 24.
a
1 2 3 4 b 5

R I

FIG. 29.   x   x

falling lines

FIG. 30.

289

Fig. 24 shows how the templet for the wreathed portion of a scroll is developed. In this case, no straight portion is required where the wreath joins the flat portion of the scroll. Notice that the tangent lines are placed on the templets. These assist in positioning the last named correctly on the timber to be shaped.

Fig. 25a shows a templet prepared and 25b how it should be placed on the plank from which the wreath is to be produced. If placed in this way it eliminates as much as possible the 'short grain' in the material. Having marked the shape of the templet on the plank its shape is cut out with the bow-saw or, better still, on the bandsaw, cutting about $\frac{1}{2}$ in. away from the lines around the curves, and about $\frac{1}{32}$ in. away from the ends. Afterwards, the ends can be cleaned up, perfectly square, with a smoothing plane.

Fig. 27 shows the wreathed portion for the scroll prepared ready for shaping. Note that the tangent lines have been placed on the timber, and these have been squared down the ends. Other lines have been squared over each end and exactly half way down the thickness of the material.

Marking the timber. The templets can now be placed on the timber, and care must be taken to position them correctly. First a sliding bevel must be set to the bevel seen in the elevation and a line marked on the wider end of the timber, allowing it to pass through the intersection of the lines squared across the end, see Fig. 28. The templets are placed on the material, one on each surface, so that the tangent lines are immediately over the ends of the bevel at one end and the tangent line on the wreath at the other end.

The next task is to remove all the timber on the wreathed portion which is outside the edges of the templets, Fig. 29. When this has been done it should be possible to place a straight-edge across the edges of the templets and have the surface of the straight-edge in contact with the shaped surface of the wreath.

The next step is to remove both templets and mark the section of the handrail on the ends of the wreath (Fig. 29). The top and bottom edges of the material are then prepared as in Fig. 30. This is done by carefully placing the falling lines on the wreathed portion in such a way as to avoid any irregularities and to assure a good sweeping curve to the handrail when completed, see Fig. 29. At all times the top and bottom edges of the wreath should be at right angles to the side surfaces.

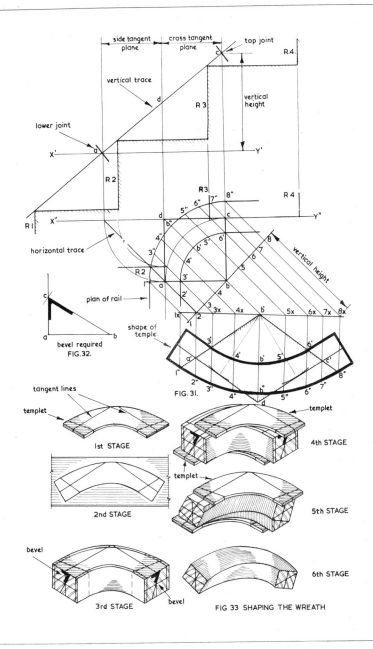

side tangent plane
cross tangent plane
top joint
R.4.
vertical trace
d
R 3
vertical height
lower joint
X'
a
Y'
R 2
R 3
R 4
X''
d
c
Y''
horizontal trace
plan of rail
R 2
vertical height
c
a
b
bevel required
FIG.32.
shape of temple
FIG. 31.

tangent lines
templet
templet
1st STAGE
4th STAGE
templet
2nd STAGE
5th STAGE
bevel
3rd STAGE
bevel
6th STAGE
FIG 33 SHAPING THE WREATH

When the wreathed portion has been prepared in its square form it is moulded and then is ready to be fitted to the straight sections of handrails by using the handrail bolt and dowelled joints.

Two-bevel work. We now come to two-bevel work in handrailing in which, as the term implies, two bevels are applied to the wreath, one at each end. It is necessary to emphasize that a knowledge of the development of oblique planes, Chapter 19, is required to follow this part of the chapter on handrailing.

Let Fig. 31 be the plan of a turn in a staircase, this being two straight flights connected by a quarter-space landing, the space between risers 2 and 3 being that occupied by the landing. Since two-bevel work is based on a prism, as in one-bevel handrailing, it is necessary to superimpose the plan of the turn on to the plan of a prism. As the two flights are at right angles to one another, the prism is square.

Let a–b–c–d be the plan of the square prism. Make the sides of the prism equal the radius of the turn to the centre line of the handrail. The centre line of the handrail travels from a round to c.

The next step is to draw the stretch-out of the steps around the turn, so with centre d and radii d–R2 and d–a in turn, project these points round to the x″–y line. The vertical line from R2 will give the position of riser 2 in the stretch-out, and vertical lines from R3 and R4 will give the positions of these risers. The stretch-out can now be drawn, making the rise of each step suit the requirements of the stairs. In Fig. 31 the distance R2–d added to the distance d–R3 are equal to the width of one tread. This has been done deliberately in this first two-bevel example to simplify the problem, as will be seen later in other examples. The falling line can be drawn through the nosings giving the positions of the joints at the edges of the side- and cross-tangent planes. These are obtained by drawing vertical lines from c, d and the point where the curve projected round from a intersects the x″–y″ line. The falling line, as it passes over the side-tangent plane, gives the vertical trace of the top surface of the prism, this being the surface we have to develop to obtain the shapes of the templets. The x′–y′ line is drawn through the point where the falling line intersects with the edge of the side tangent plane.

Reference should now be made to the oblique planes in Chapter 19 to see how the surface a–b–c–d is developed, Fig. 58. Having obtained this shape, a′–b′–c′–d′, one can proceed to develop the shape of

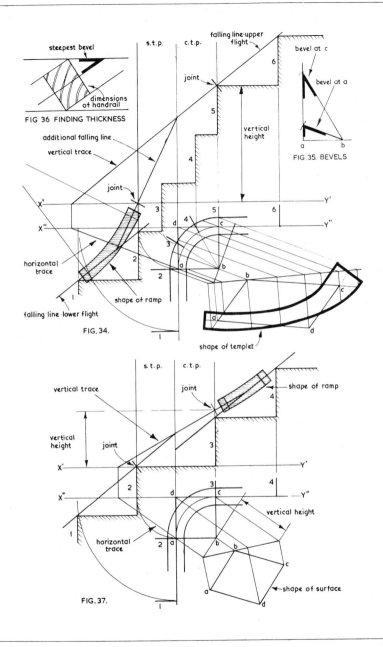

steepest bevel

dimensions of handrail

FIG 36 FINDING THICKNESS

s.t.p.

c.t.p.

falling line·upper flight

joint

bevel at c

bevel at a

FIG.35. BEVELS

6

5

vertical height

additional falling line

vertical trace

joint

4

X'

Y'

X''

Y''

3

d 4

c

3

horizontal trace

2

a

b

b

c

shape of ramp

falling line·lower flight

a

d

FIG.34.

shape of templet

s.t.p.

c.t.p.

vertical trace

joint

shape of ramp

4

vertical height

joint

3

X'

Y'

2

X''

Y''

d

3

c

4

vertical height

horizontal trace

2

a

b

b

c

shape of surface

a

d

FIG.37.

293

the templets. To do this the outline of the curved portion of handrail must be placed on the plan of the turn. The distance from 1–8 should be divided into a number of parts, and lines drawn from these points across the plan parallel to the horizontal trace. They should then be projected down to the 1x–8x line and from this line downwards at right angles to 1x–8x. The lengths of the lines running across the development of the surface a–b–c–d should then be made to equal those across the plan of the turn to give points on the shape of the templets. Straight sections about 2–2½ in. long are added to each end of the templet development, these straight sections being parallel to the tangent lines a′–d and c′–d. The ends of the templet must be at right angles to the tangents a′–d and c′–d.

Bevels. To obtain the bevels required to apply to the ends of the wreathed portion, first construct a right angle, Fig. 32, making a–b equal to a–b in the plan. Next place the compass point in b′ in the development of the top surface, and open them to just touch tangent a′–d. Next place the compass point in b on the right angle, Fig. 32, and cut the vertical arm in c. This will give the bevel to apply to lower end of the wreath, near a, in fact. The other bevel is obtained in the same way. With compass point in b′ open the compasses to just touch tangent c′–d, and place the compass point in b Fig. 32 and cut the vertical arm of the right angle. It will be found in this case, that the vertical arm will be cut in c, making the bevel for the top end of the wreath the same as that for the lower end. This is not always the case, as will be seen later. Fig. 33 shows the various stages in shaping the wreath.

Placing the templets. The only difference in these drawings is the positioning of the templets. As there is a bevel at each end of the wreath, care must be taken to see that the tangent lines on the templets are immediately over the lines running from the top and bottom ends of the bevels, and parallel to the tangents on the wreath (see 4th stage).

Fig. 34 is the development of the shape of the templets for a staircase of two straight flights connected by a quarter-space of winders. The stretch-out of the steps and the development of the surface a–b–c–d is straightforward if note has been made of the last example and the geometry of oblique planes.

There is, however, a complication in this and similar examples. The falling lines of the lower and upper flights do not meet, and as it is necessary for them to meet to obtain a continuous handrail round the turn, it is essential to introduce another falling line so that this will connect the other two. This also means introducing a ramped portion of handrail in between the wreath and the lower straight handrail, as seen in Fig. 34.

The top falling line should be brought down to the vertical line immediately above d in the plan, and from here the additional falling line placed in to meet the lower falling line near riser 1. This will allow the ramp to be included—its end to be some distance, say 2 in., away from the lower joint line of the wreath, which, incidentally, must fall on line x′–y′ and the edge of the side-tangent plane. The top joint, of course, is on the edge of the cross-tangent plane.

The vertical trace is the continuance of the top falling line, and should extend down to the x′–y′ line, which passes through the lower joint, from here vertically down to the x″–y″ line to obtain the position of the horizontal trace.

Two bevels are required, as seen in Fig. 35, and these are developed in the way already explained. The thickness of material is found by taking the steeper of the two bevels and proceeding as before, see Fig. 36.

Fig. 37 is another staircase of two flights connected, this time by a quarter-space landing. The prism plan a–b–c–d can be seen, and the stretch-out of the steps shows that the two falling lines, again, do not meet. This time, however, the lower falling line is above the position of the upper falling line, and so it is necessary to introduce a ramp at the upper end of the curved portion of handrail. Having ascertained the position and shape of the ramped portion and the development of the templets, the bevels and thickness of material required can proceed as before.

The next problem, involving a turn in a staircase which is not a right angle (see Fig. 38), also necessitates the introduction of a ramp. The prism on which this problem is based is trapezoidal in plan. The staircase consists of two straight flights connected by winders with the risers numbered 2, 3 and 4. Again, the falling lines do not meet and a ramp must be used, this time at the lower end of the curved portion. A study of this type of oblique surface, Fig. 60, Chapter 19, will assist the reader in understanding the method used for developing the surface a–b–c–d and also the shape of the templets. The bevels are shown in Fig. 39, and are developed as before.

Some readers may find it difficult to understand the geometry used for developing the oblique surfaces in the problems on hand-railing. Fig. 40 has been included in this chapter for guidance. First try to imagine that the prism (a square one in this case) has been placed in the angle set up by the vertical and horizontal planes, Fig. 40. a–b–c–d is the top surface which has to be developed. The edge c–d is projected down to the x–y line and this projected line is the vertical trace of the top surface. Another line from where the vertical line meets the x–y line and passing through the lowest corner of the top surface will give the horizontal trace. We must now try to imagine a right-angled triangle, a'–c–c'', standing at right angles to the horizontal trace and in contact with the corner of the prism of which b is the top point. The edge c–c'' is equal to the vertical height between the lowest and highest corners of the prism—a and c.

It should also be noted that all the lines which start from the top edge of the triangle and go across the top surface of the prism are parallel to the horizontal trace, and are therefore all horizontal lines and can be measured in the plan.

If the triangle a'–c–c'' were turned so that it lay on the horizontal plane as seen in Fig. 40 and the various lines projected downwards at right angles to the line a'–c' (which is now the top edge of the triangle), and all those lines made equal in length to those across the top surface of the prism, not only can the shape of the top surface be drawn but the curve a–c can also be plotted.

Dancing steps. Before leaving the subject of staircase work it will be well to consider a stair with dancing steps, Fig. 41. This staircase must not be confused with those in publications illustrating the new Building Regulations, where certain requirements are now in existence regarding winders in staircases. Readers must refer to these to ascertain what is actually required. To overcome the danger of very narrow steps near the newel post at the change of direction, it is possible to increase the widths of the winders by introducing additional tapered steps (dancing steps). From the plan Fig. 41 it will be seen that there are six of these instead of the usual three at the turn. The walking line of a staircase is the path of an individual ascending or descending the stair and this is considered to be approximately 16 in. from the outer string. When possible, it is wise to have the going of all the steps of a stair equal at the walking line, and so this should be established before

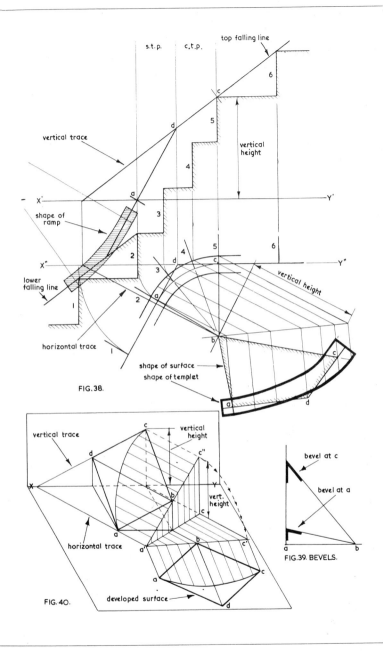

FIG. 38.

shape of ramp

lower falling line

horizontal trace

vertical trace

s.t.p.   c.t.p.   top falling line

vertical height

shape of surface
shape of templet

FIG. 39. BEVELS.

bevel at c

bevel at a

vertical trace

vertical height

vert. height

horizontal trace

developed surface

FIG. 40.

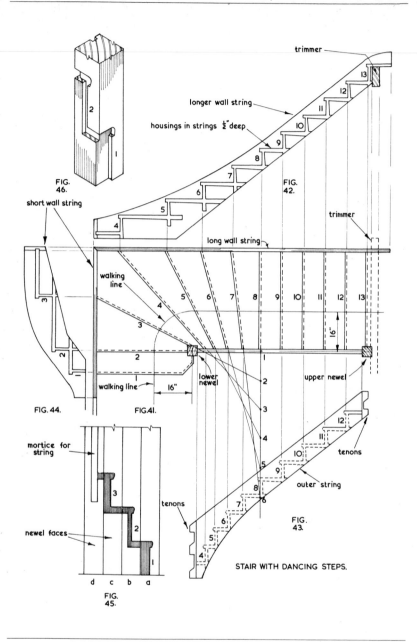

FIG. 46.

short wall string

longer wall string

housings in strings ½" deep

FIG. 42.

trimmer

13
12
11
10
9
8
7
6
5
4

long wall string

trimmer

walking line

5   6   7   8   9   10   11   12   13

4

3

16"

2   1

lower newel   upper newel

walking line

1   16"

FIG. 44.   FIG. 41.

mortice for string

3

2

newel faces

1

d   c   b   a

FIG. 45.

2
3
4
5
6
7
8
9
10
11
12
13

1

tenons

outer string

tenons

FIG. 43.

STAIR WITH DANCING STEPS.

setting out the positions of the risers. Having drawn the walking line on the plan and marked the 'going' on the walking line, it is necessary to decide on the number of dancing steps one is prepared to include in the flight.

To place on the drawing the positions of the risers to the dancing steps, draw a vertical line down from the first riser which is at right angles to the strings, in this case riser number 8, and mark off on this line from the centre of the outer string a number of spaces, say 12 in. apart. The number of spaces should be equal to the number of dancing or tapered steps. From these points lines can be drawn through the points on the walking line to obtain the positions of the risers to the tapered steps. It may be necessary to adjust the distances 1–2–3 etc. until a satisfactory positioning of the winder risers has been obtained.

Also shown on the drawing are methods for obtaining the shapes of the strings and the positions of the step housings, and it will be seen that the outer string has to be increased in width considerably towards its lower end because the falling line is much steeper at this point. This is due to the fact that the width of each step from number 8 downwards is narrower at the outer string than it is where it enters the long wall string. Fig. 45 shows how the four faces of the lower newel are set out for recessing, and Fig. 46 is a pictorial view of the newel. Although this type of staircase produces one which is safe and easy going, one must also remember that it is more expensive to produce.

32 Glulam Work

The word 'Glulam' is a term used to describe a method of building up a timber component of any shape to any requested dimensions, and involves gluing together a number of laminations on a specially prepared cramping area. Although this has become a highly specialized section of the woodworking industry in recent years, a brief description of the procedure used in producing a glulam item will help the woodworker should he be called upon to assist in this type of production.

Small work. For small pieces of curved work made in a similar manner to that to be described in these pages, no special preparation is necessary. Fig. 1 shows how small items of work can be manufactured. There is a flat base board, say of 1 in. blockboard or plywood, strengthened if necessary by battens glued and screwed to the underside. The double outline of the item to be produced is marked carefully on the surface, and formers, cut to the shape are screwed on to coincide with the inside of the two lines.

Glue is spread on both surfaces of the laminations except the two outside pieces which should have only one surface glued. The pieces are loosely placed together on to the base board, and a second set of formers, cut to coincide with the outer outline of the item to be produced, are placed behind the laminations. By using a number of G cramps across the sets of formers the work can be cramped up to bring the surfaces of all the laminations into close contact, see Fig. 1.

Large work. For structural laminated work much greater care must be taken. For instance, the selection of the right grade of timber is important. There are two groups of softwood timbers which are recommended for structural purposes, the first including Douglas fir, longleaf pitch pine, and shortleaf pitch pine. The second group

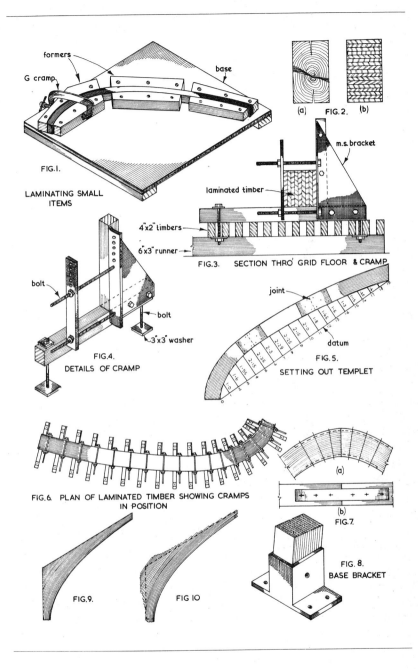

formers

G cramp

base

FIG.1.

LAMINATING SMALL
ITEMS

(a) FIG. 2. (b)

m.s. bracket

laminated timber

4"x2" timbers

6"x3" runner

FIG.3. SECTION THRO' GRID FLOOR & CRAMP

bolt

bolt

3"x3" washer

FIG.4.
DETAILS OF CRAMP

joint

datum

FIG. 5.
SETTING OUT TEMPLET

FIG.6. PLAN OF LAMINATED TIMBER SHOWING CRAMPS
IN POSITION

(a)

(b)
FIG.7.

FIG.9.

FIG 10

FIG. 8.
BASE BRACKET

includes Canadian spruce, European larch, redwood, whitewood, and western hemlock. More information on these recommended timbers for structural work can be obtained from B.S.1860: 1952 *Structural Softwood—Measurement of Characteristics Affecting Strength,* and the British Code of Practice C.P. 112 (1952) *The Structural Use of Timber in Buildings.* These, in addition to gluing tables of recommended stresses for the two groups of timbers, also set out the effect of various degrades or defects on the timbers. Maximum allowable sizes of these defects are given to assist those whose job it is to select timber for structural work.

Timber suitability. Among the characteristics one has to consider when grading softwoods is the slope of the grain. This can vary from 1 in 8 for beams and compression members not more than 4 in. thick, to 1 in 11 for compression members more than 4 in. thick. Wane, too,

A view of laminated arches to a large dance hall. *CIBA (A.R.L.) Ltd.*

must be considered and so must the rate of growth, because softwoods derive their strength from the tracheids of the latewood or summer growth. A slow rate of growth usually indicates a strong timber. Knots, shakes and splits affect the strength of a piece of timber and so these have to be measured and taken into calculation.

One advantage of using timber for glulam work is that, since it is formed from laminations, defects such as knots can be dispersed more evenly throughout the finished product. Fig. 2 shows the difference between a solid beam a, and a laminated beam, b. The solid piece can be much weaker than the laminated one because serious defects can be concentrated at one or more points.

Many firms now producing glulam components have their own methods for producing such work, and most install what they consider to be the best kinds of cramps for the work they are producing. Fundamentally, however, all systems are much the same. Many of the cramps or chairs, as they are often called, are made from metal angles and channels but basically they are similar to that shown in Fig. 4. This consists of two pieces of timber jointed to form a right angle with a mild steel bracket bolted to them. The bracket should be made from mild steel plate $\frac{3}{8}$ in. thick, with one of its edges turned at right angles to the main portion which is bolted to the timbers. In this returned edge are drilled a series of holes through which bolts

A large glulam beam on the cramping bed. *Leicester, Lovell & Co., Ltd.*

pass. The other edge of the cramp consists of a piece of $\frac{3}{8}$ in. mild steel plate with bolt holes drilled through it to coincide with those in the bracket. Two more bolts with large square washers are provided to enable the cramp or chair to be fixed in position on the cramping area.

Cramping area. This should be made so that cramps can be fixed down on to the surface in any desired position. One method consists of a floor of slats fixed to runners such as that in Fig. 3. Runners 6 in. by 3 in. are secured to the concrete or timber floor, and 4 in. by 2 in. or 4 in. by 3 in. timbers fixed to the runners, with spaces between, to form the floor of the cramping area. The drawing also shows a cramp fixed to the slatted floor by means of the two bolts passing through the slats with large square washers below.

Fig. 6 shows the plan of a number of these cramps placed to produce a glulam component in the shape of a parabolic curve. To produce this it is necessary to set out the curve, full size, in the workshop so that the cramps can be positioned correctly. If it were considered necessary to produce a templet to assist in the cramp positioning, one method for producing it is seen in Fig. 5. This consists of drawing lateral lines at right angles to a datum line at regular intervals, and scaling the lengths of the laterals from the architect's or designer's drawing.

In Fig. 5 it can be seen that the datum is 19 ft. long and perpendicular laterals have been constructed every 12 in. along its length. These have been carefully measured on the drawing and transferred, full size, to the workshop setting-out. The templet, when made, can be used for positioning the cramps and for trimming the component to shape when the gluing up has been completed.

Assembling. When the cramps have been fixed down on to the cramping area, the building up of the component can commence. For an item such as that shown in Fig. 6 laminations approximately $\frac{3}{8}$ in. to $\frac{1}{2}$ in., according to the shape, would be used. Up to six of these would be assembled at a time, these being left until the next day when a similar number would be added. This procedure would be repeated every twenty-four hours until the component had been built up to its required dimensions.

It would then be manhandled or lifted by mechanical means to a convenient position for trimming with a portable saw and cleaning up to size and finish with a portable electric plane and sander. Lastly it would be varnished or sealed, whichever was appropriate, and any additional work carried out before delivery to the site.

Figs. 9 and 10 show an item before and after finishing. Fig. 10 illustrates how the part would appear after removal from the cramps, and Fig. 9 how it would look when completed. It should be obvious from these drawings that a templet is necessary to bring the item to the required shape.

Figs. 7 and 8 show two methods of fixing for glulam work. Fig. 7 has a mild steel strap for holding two halves of an arch at the crown, and Fig. 8 a mild steel base bracket for fixing one of its feet.

Surfaces to be glued should be planned to fairly accurate dimensions. Glue can be spread by a brush on a small job, but for the larger type of work, such as roof arches and the like, a mechanical glue spreader is desirable.

A glulam member in the course of assembly. *CIBA (A.R.L.) Ltd.*

A building, incorporating laminated arches and beams in the course of construction. Note that the smaller cross members which are secured to the beams and arches are fixed at each end by Trip–L–Grip anchors. *Kingston (Architectural Craftsmen) Ltd.*

Adhesives. In recent years much advancement has been made in the production of adhesives for the building industry. Animal and casein glues have been replaced by synthetic resins, the latter being far superior for glulam work. They are proof against dampness and are not attacked by fungi and other organisms. For this type of work they should be gap filling. Probably the best glues are resorcinol-formaldehyde and ureaformaldehyde. Two methods are used. In the first the resin is spread over one of the surfaces and the hardener on the other, the setting commencing when the two surfaces are brought in contact. In the second method the resin and hardener are mixed and spread in one application.

A glue spreading machine. This is valuable when large surfaces have to be glued quickly such as in glulam work.

33 Applied Geometry

Also see chapter 5.

A large percentage of the work of the carpenter and joiner, in the advanced stages at least, involves geometry, and it is the purpose of this chapter to give the craftsman an idea of how a knowledge of geometry can be applied to practical work.

When straight and curved lines join, as in the case of mouldings and, as Fig. 7 shows, in the construction of templets, etc., it is necessary to ensure that no irregularities occur at these points. Therefore the straight line which joins a curved line must always be tangential to the curve. The first six illustrations explain how internal and external tangents to curves can be constructed, and Fig. 7 shows how this knowledge can be applied in a practical manner.

Tangents. Fig. 1 gives the method of drawing a tangent to a circle from any point p outside the circle. Draw the circle and mark point p in the desired position. Join p to the centre of the circle, a, and bisect p–a in b. With compass point in b and radius b–p describe a semi-circle to cut the circle in c. Point c is the point of contact between the circle and the tangent.

Fig. 2 shows how to draw an external tangent to two unequal and touching circles. Draw the circles, any diameter, so that they touch in d. Join the centres a and b with a straight line and bisect a–b in c. With centre c and radius c–a draw the semicircle a–e–b. From d draw d–e at right angles to a–b to give point e on the semicircle. With centre e and radius e–d describe the semicircle f–d–g. Points g and f are the points of contact for the tangent g–e–f. g–a and f–b are normals and are at right angles to the tangent.

Fig. 3 shows how to draw an external tangent to two unequal circles which are some distance apart. Draw the circles and a line to join their centres a and b. Bisect a–b to give c, and with centre c and radius

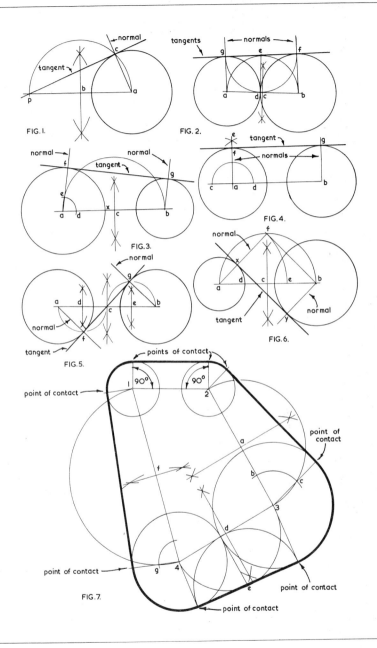

FIG. I.

FIG. 2.

FIG. 3.

FIG. 4.

FIG. 5.

FIG. 6.

FIG. 7.

c–a draw the semicircle. From x, on the larger of the two circles, make x–d equal to the radius of the small circle. With compass point in a, and radius a–d describe a curve to cut the semicircle in e. Draw a line from a to pass through e and give point f on the circumference of the large circle. Add a line from b, parallel to a–f to give point g on the small circle. Points g and f are the points of contact.

To draw an external tangent to two equal circles some distance apart, Fig. 4, describe the circles and join their centres a and b with a straight line. Draw a line at a perpendicular to a–b to give point f on the circumference of that circle. Another line from b, perpendicular to a–b (which will be parallel to a–f) gives point g on the second circle. Points f and g are points of contact for the tangent.

To construct an internal tangent (one which passes between the circles) to two equal circles some distance apart, describe the circles and join their centres a and b, Fig. 5. Bisect a–b in c, and bisect a–c and c–b to give points d and e. Use d and e as centres to draw the semi-circles a–f–c and b–g–c. Points f and g are the points of contact.

Finally, to work out the internal tangent to two unequal circles, Fig. 6, draw the two circles some distance apart and join their centres a and b. Bisect a–b in c and draw the semi-circle a–f–b. Make d–e equal to the radius of the larger circle. With compass point in a and radius a–e draw the arc to cut the semicircle in f. Draw straight lines from f to a and b, and also a line from b perpendicular to f–b to give point y. This is one point of contact, the other being x.

Templet with joining straight and curved lines. Fig. 7 shows how an irregular figure (which might be the shape of a templet) involving some of the tangent problems in Figs. 1–6 can be tackled.

Scroll problems. Other problems involving parts of circles which have not been mentioned in the companion volume *(Practical Carpentry and Joinery)*, are those concerned with spirals and scrolls. Fig. 8 shows how an Archimedean spiral can be set out. It could be adapted for use at the end of a straight wall handrail as seen in Fig. 9. Draw the large circle into which the spiral is to be constructed, and divide it into any number of parts, say twelve, and number these as shown on the drawing. From point 1 and along the diameter line mark off the same number of equal spaces, making the overall distance 1–13 the depth of the handrail. Using the centre of the circle and opening the compass to the various points on the diameter line,

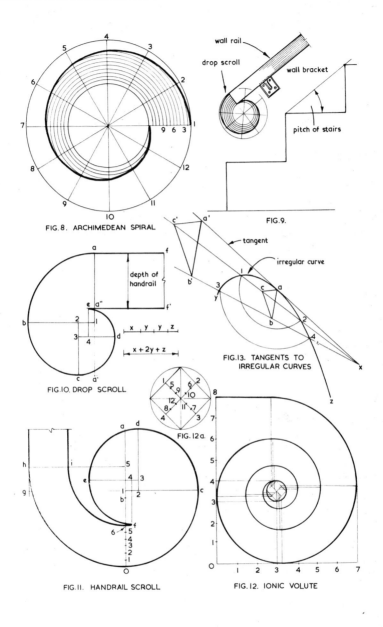

FIG. 8. ARCHIMEDEAN SPIRAL

wall rail
drop scroll
wall bracket
pitch of stairs

FIG. 9.

depth of handrail

x · y · y · z

x + 2y + z

FIG. 10. DROP SCROLL

tangent
irregular curve

FIG. 13. TANGENTS TO IRREGULAR CURVES

FIG. 12 a.

FIG. 11. HANDRAIL SCROLL

FIG. 12. IONIC VOLUTE

positions on the spiral can be obtained by swinging the various arcs round to the appropriate normal, as seen in the drawing. A freehand curve through these points will give the required spiral. If this type of scroll is used for the wall handrail in Fig. 9 a slight adjustment will have to be made on the curve from, say, normal number 11 so that the scroll portion will not interfere with the straight rail.

Fig. 10 shows how a drop scroll can be set out to a pleasing line. This is constructed with quadrants of circles. Let a–a′ be the overall depth of the scroll, and a–a″ the depth of the handrail. Draw the two horizontal lines a–f and a″–f′. Next, draw a line equal to a–a″ in length, and divide it up into sections equal to x + 2y + z, as seen in the drawing, ensuring that x is larger than y and y is larger than z.

On line a–a′ and some distance below a″ mark off centre number 1. The position of this centre will determine the size of the completed scroll, and it may therefore be necessary later to adjust the position of this centre if the completed scroll proves to be too large or too small.

With centre 1 and radius 1–a describe the quadrant a–b. Along the line 1–b mark off the distance x to give point 2, and describe the quadrant b–c; on line 2–c mark off distance y to give point 3 and describe quadrant c–d; on line 3–d mark off distance z to give the remaining centre to be used to complete the scroll. If the work is accurate the end of the scroll will terminate at e, which is on the line a″–f′.

Scroll to geometrical stair. Fig. 11 shows how to set out a handrail scroll at the lower end of a geometrical staircase. Make O–a equal to four-fifths of the overall width of the scroll, and O–f equal to the width of the handrail. Divide O–f into six equal parts. Bisect O–a in b and step off towards a, one-sixth of the width of O–f. This is centre number 1, and it should be used for describing the quadrant O–c. On line 1–c mark off two-sixths of the width of O–f to give centre 2 and, using this point, draw quadrant c–d. On line 2–d mark off $1\frac{1}{2}$ sixths O–f to give centre 3 for quadrant d–e, and on line 3–e mark off one-sixth O–f to give centre 4. The quadrant from e should terminate opposite point 6 and directly below centre number 4. Make c–g equal to the width of the scroll, and with radius 1–g place the compass point in O and mark off the centre 5 on O–a. With centre 5 describe quadrants O–h and 6–i to complete the scroll.

Ionic volute. Fig. 12 is an Ionic volute, and fits into a rectangle eight units high and seven units wide. Its centre eye is one unit in diameter and is placed between points 3 and 4 on the vertical arm with its centre immediately over point number 3 on the horizontal arm. Fig. 12a shows how the centre eye is set out and numbered, each being used in numerical order to construct a quadrant. Accuracy is most important in this as in other scrolls if a pleasing result is required.

Irregular curve tangent. Fig. 13 shows how to draw a tangent to an irregular curve. Draw the curve y–z any shape, and assume that a tangent is required at point a. With centre a draw two arcs to give points 1 and 2, 3 and 4 on the curve. Draw a straight line through points 1 and 2 and another through points 3 and 4 to meet the first line in x. From x draw the tangent through a.

If point x is inaccessible, any triangle is constructed at a to give points b and c on the lines passing through the curve, and then another triangle, similar to the first (all sides parallel to those of the first triangle) to obtain point a´. Draw a line from a´ through a to obtain the tangent.

Solid geometry. Let us now turn our attention to solid geometry. The study of geometrical solids such as prisms, cylinders, pyramids, etc., is essential to the carpenter and joiner because most of the shapes found in the woodworking industry can be related to one or more of these solids. The basic developments of the common geometrical solids were dealt with in the companion volume, *Practical Carpentry and Joinery*. Here it is the aim to show how these developments can be applied to practical work.

Rectangular sections. Let us take rectangular prisms, or rectangular pieces of timber first. A simple case could be a square post supported by two square struts set at an angle of 60° (Fig. 14). The strut to the left shows that it has one of its surfaces facing directly upwards, and that to the right has one of its corners pointing upwards. The problem here is to produce the bevels so that the struts can be cut to fit up against the post.

All that has to be done to produce the bevels for the strut on the left is to develop its top surface. Draw a horizontal line out from point b in the elevation, and with centre b and radius b–a describe an arc to give a´ on the horizontal line. Drop a line from a´ vertically downwards to give point a´ on the horizontal line brought out from a in the plan. Join c to a´ and b to a´ to give the bevels to apply to the top and bottom surfaces of the strut. The bevel to apply to the front and rear surfaces is seen in the elevation.

To develop the bevels for the strut on the right of the drawing, the two sides of the strut nearest the front must be developed. Draw the plan and elevation as shown, and place on the elevation a section of the strut. Draw a line through x on the section at right angles to the corners. With compass point in x and radii x–x′ and x–x″ describe arcs to give points on the line which passes through x. From these points draw the edges of the developed sides. From points e, f, h and i draw lines outwards to intersect with the developed edges in e′, f′, h′ and i′. Join d to e′, d to i′, g to f′ and g to h′ to represent the developed sides and the required bevels.

Fig. 15 shows the plan and elevation of another post supported by a strut, the strut situated towards the front of the post. Draw the plan and elevation and put in a line above and at right angles to the strut, and step off four distances equal to the width of the sides of the strut. From a, b, c and d draw lines at right angles to the inclination of the strut to give points a′, b′, c′ and d′. The figure thus produced will give the development of the four surfaces and also the required bevels.

Hexagonal prism. Fig. 16 shows a small hexagonal prism passing through a larger square prism. The problems are to develop the portion of the smaller prism which projects from one corner of the other, and to develop the hole in the large prism through which the smaller one passes. Draw the plan and elevation. Project lines from a′, b′, c′, x, y and the end of the small prism downwards and at right angles to its elevation. On the line from the top end of the small prism, and starting from any convenient point, step off six distances equal to the widths of the sides of the small prism. From these points draw lines parallel to the small prism to intersect with the lines from a′, b′, etc., to give points a′, b′ and c′, etc. on the development. The positions of x and y are halfway across the sides in which they are situated. To develop the hole near the top of the square prism, project over to the right, and at right angles to the large prism, x, a′, b′, c′ and y, and starting from a centre line x–y, step off to the left and to the right the distances 1 and 2 seen in the plan. Vertical lines through these points will give the required points on the hole development.

Pyramids. We now come to the question of pyramids. Fig. 17 is the plan and elevation of a square box with inclined sides. Two corners

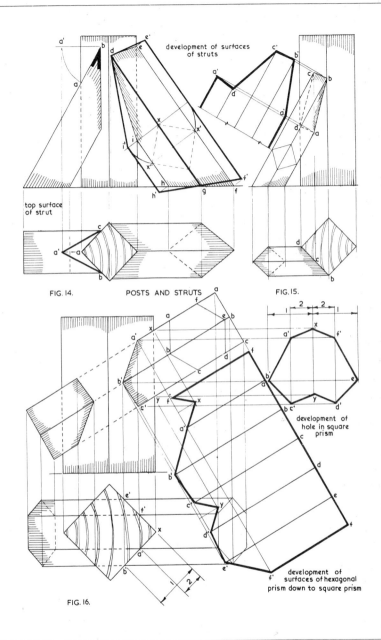

development of surfaces
of struts

top surface
of strut

FIG. 14.   POSTS AND STRUTS   FIG. 15.

development of
hole in square
prism

development of
surfaces of hexagonal
prism down to square prism

FIG. 16.

are butt-jointed, and the other two mitred. To develop the shapes of the sides of the box, with compass point in f in the elevation and radius f–d, describe an arc to give d on the extended base line. Drop a vertical line from d to give points h′ and d′ on horizontal lines brought out from h and d in the plan, f–d′–h′–g being the development.

To obtain the bevel to apply to the butt-jointed ends, draw a right angle at d and make d–e equal to the vertical height of the box. Join e to f, e–f being the true length of the corner d–f. Draw line 5–6 at any point but at right angles to d–f, and with compass point in 7 and opened to just touch line f–e describe an arc to give point 8 on d–f. Join 5 to 8 and 6 to 8. Angle 5–8–6 is the dihedral angle set up by the two sides, and angle 5–8–9 the bevel to which the ends of the sides are cut to so that the butt joints are perfect fits. The mitre bevel for the two other corners is found in the same way, but this time half of the dihedral angle is taken for the bevel.

Fig. 18 shows the top corner of some splayed linings round a window or door opening. The problems are to develop the shape of the linings where they intersect, and the bevel so that they can be butt-jointed or tongued-and-grooved. First draw the elevation of the corner and a section through one of the linings. With compass point in a in the plan, and radius a–b, describe an arc to give b′ on the horizontal line passing through a. Project upwards vertically a line from b′ to give b′ on the horizontal line brought out from b in the elevation. a–b′–c′–d is the shape of the ends of the linings where they intersect.

The development of the bevel to apply to the ends so that they can be butt-jointed or tongued-and-grooved is shown on the drawing, and is similar to the bevel for butt-jointing the corners of the box in Fig. 17. Distance y is equal to y in the plan.

Octagonal roof. Fig. 19 shows how the surfaces of a roof to an octagonal shed or kiosk can be developed. The drawings are fairly straightforward and similar to those in Fig. 17 and should need no description. The backing bevel, which is applied to the top surfaces of the hips so that a seating is provided for the tile battens, is found as for the mitre bevel for the box corners.

Oblique pyramid. Fig. 20 shows how the sides of an oblique pyramid should be developed. Remember that with a right square pyramid all four corners are equal in length, but with the oblique pyramid two

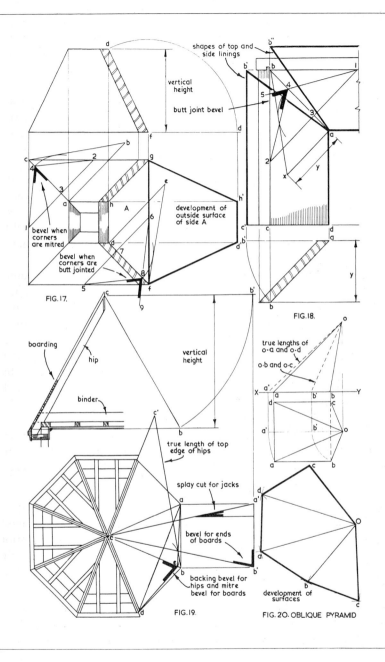

shapes of top and side linings

vertical height

butt joint bevel

development of outside surface of side A

bevel when corners are mitred

bevel when corners are butt jointed

FIG. 17.

FIG.18.

boarding

hip

binder

vertical height

true length of top edge of hips

splay cut for jacks

bevel for ends of boards

backing bevel for hips and mitre bevel for boards

FIG.19.

true lengths of o-a and o-d o-b and o-c.

development of surfaces

FIG. 20. OBLIQUE PYRAMID

of the corners are longer than the other two. The plan and elevation show how the lengths of the corners are developed. Having obtained these lengths the sides are developed as follows. Draw the corner o–c, making this equal to the development in the elevation. With compass open c–d in the plan and with compass point in c in the development, make an arc in the direction of d. With compass open the developed length of o–d and with its point in o in the development, make an arc to cut the first arc in d. The other points in the development can be obtained similarly.

Cylinders. Fig. 21 shows a section through a pipe or cylinder passing through a partition at an angle of 45°. The problems are to develop the shape of a clearance hole in the partition, and to develop the shape of the surface of the pipe section which passes through the partition.

The pipe and partition should first be drawn as shown, and a section of the pipe placed over that part which passes through the partition. Another circle, to represent the section of the hole in the partition, should also be placed on the drawing, making the difference in the diameter of the circles equal the amount of clearance required between the pipe and the hole surface. Divide these sections into, say, twelve equal parts.

To develop the shape of the hole, project the points on the outside circle up to the surface of the partition, and from here over horizontally to where the shape is to be developed. Draw the centre line of the development, 1′–7′, in a vertical direction, and make the distances 6′–8′, 5′–9′, etc. equal to those of the outside circle representing the hole in the partition. A freehand curve through the points obtained will give the shape of the hole to be marked on the partition surface. The shape is elliptical.

The shape of the pipe surface can be developed thus. Project to each end of the pipe the various points 1, 2, 3, etc. on the circle representing the pipe section, and draw lines from these points downwards and at right angles to the pitch of the pipe. On one of these lines and from any convenient point mark off twelve distances equal to those round the pipe. Draw lines from these points across the two sets of lines brought over from the two ends of the pipe to give two sets of intersections through which can be drawn freehand curves to obtain the shape of the pipe surface.

Fig. 22 shows a similar problem; how to obtain the shape of the

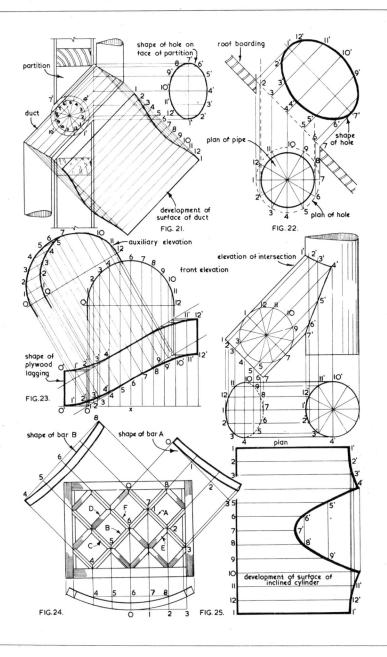

shape of hole on face of partition

partition

duct

development of surface of duct

FIG. 21.

roof boarding

plan of pipe

shape of hole

plan of hole

FIG. 22.

auxiliary elevation

front elevation

elevation of intersection

shape of plywood lagging

FIG. 23.

plan

shape of bar B

shape of bar A

FIG. 24.

development of surface of inclined cylinder

FIG. 25.

hole in the boarding on a roof to allow a round flue pipe to pass through. Again, the pipe should be given a clearance hole and a study of the drawings should explain how the shape of the hole is obtained.

Fig. 23 shows how the plywood lagging to a centre for a skew arch is developed. The arch, semicircular when viewed parallel with its jambs, should be divided up into, say, twelve parts, and these points projected down to each edge of the plan. On the horizontal line below the plan should be marked twelve spaces equal to those round the elevation curve, x being the centre point. Project all these points upwards vertically to intersect horizontal lines brought out from the points on each side of the plan, thus giving points on the development of the soffit of the arch, which is also the shape of the lagging required.

Fig. 24 is the plan and elevation of a sash, circular in plan with diagonal glazing bars. To obtain the shape of glazing bar A, project points 0, 1, 2 and 3 downwards to the horizontal line just below the plan, giving points 0, 1, 2 and 3 on this line. Also project lines from 0, 1, 2 and 3 over and at right angles to the bar in the elevation, and draw the line 0–3 parallel to the bar in the elevation. Make the distances between the line 0–3 and the development of the bar equal to those seen in the plan. This shape can then be used as a templet for the bars marked A, C, D and E. The shape of the templet for bars B and F is found similarly as shown in the drawings.

Intersecting cylinders. Another exercise involving the cylinder which often proves valuable is shown in Fig. 25. The top drawings show the plan and elevation of two intersecting cylinders. The problem here is to develop the shape of the surface of the inclined cylinder so that it will fit exactly over the vertical one.

Draw the plan and elevation and divide the surface of each cylinder into, say, twelve equal parts. Project the various points on the plan of the inclined cylinder over to the plan of the vertical one to give points $1'$, $2'$, $3'$, etc. From here they should be projected upwards to intersect with the lines taken along the elevation of the inclined cylinder to give $1'$, $2'$, $3'$, etc. A freehand curve through the points thus obtained will give the elevation of the intersection.

To develop the surface of the inclined cylinder, draw the vertical line 1–1 seen below the plan, and mark off twelve distances equal to those round the surface of the inclined cylinder. Draw horizontal lines from these points and make these lines equal to those on the surface. These lengths can be obtained from the elevation. Freehand curves through the ends of the lines will give the surface development.

Cone problems. Fig. 26 is the plan and elevation of a cone. In the companion volume *(Practical Carpentry and Joinery)*, it was shown how to develop the surface of the solid, and also how to develop an elliptical section through the cone. Fig. 26 explains the method of developing two more sections, namely, the parabola and hyperbola. If the section line is parallel to one of the edges of the elevation, x–7′, a parabolic section will be produced, when the line is vertical, but not on the centre line of the elevation, a–7″, a hyperbolic curve will be the shape of the section.

To produce the parabolic section, draw the section line parallel to one edge of the elevation, and drop lines down vertically from points 3′, 4′, etc. to give a plan of the section. Then draw lines at right angles to the section line in the elevation from all the points on line x–7′, and, starting from a centre line z–7′, make the distances across the development equal those across the plan of the section.

To produce the hyperbolic section, draw line a–7″, being careful not to position it on the centre line of the solid. Project the line down into the plan to obtain the plan of the section. This is also a straight line. From the various points on the plan project lines horizontally over to where the section is to be produced, and on the centre line c–7″ mark off the various heights taken from the elevation; for example, make c–d in the development equal a–5″ in the elevation, c–e equal a–6″, and c–7″ equal a–7″. Project lines parallel to the base of the development to give 5″, 6″, 8″ and 9″. A freehand curve through these points will give the development of the hyperbolic section.

Parabola. Figs. 27 and 29 show two more methods which can be used for setting out the parabola. When the height and the width are known the method shown in Fig. 27 can be used; when the positions of the directrix and the focal point are known the method shown in Fig. 29 is used. When using this second method it must be remembered that the distances between the directrix and the curve, and the focal point and the curve is always in the ratio of 1:1.

If it is necessary to find the position of the focal point of a parabola the method shown in Fig. 28 should be followed. Construct the parabola as in Fig. 27 and construct a tangent at any point p. Mark point p anywhere on the curve and draw a horizontal line from it over to the centre line. With compass point in b and radius a–b draw the semicircle to give a′ on the centre line. Draw a line from a′ to p. This is a tangent to the curve at p. To produce the focal point draw a vertical line upwards from p, and, with p as centre and radius any convenient distance, draw the arc d–g, cutting the tangent in e. With compass point in e and radius e–d draw an arc to give point g on d–g. Draw a line from g to p, cutting the centre line in F. This is the focal point of the parabola.

If a number of normals are required from various points on the parabolic curve, say, from p′ and p′′, place the compass point in F and with radii F–p′ and F–p′′ draw arcs to give x′ and x′′ on the centre line. Lines through x′–p′ and x′′–p′′ will give the required normals.

Pitched and conical roof intersections. Fig. 30 is the plan and elevation of a conical roof intersecting a pitched roof. The main roof is pitched at the same angle as the rafters of the conical roof. The problem is to develop the shape of the curb required to be placed on the main roof so that the ends of the rafters on one half of the conical roof will have a seating, and to which they can be fixed. This problem is similar to producing a section through a cone.

The section being parallel with the edge of the cone, the shape of the section (or curb, as it is in this case) is parabolic. Compare this drawing with Fig. 26.

Fig. 31 is the plan of the roof of a circular bay window abutting the wall of a house. The width of the roof from left to right is x–y, and the distance it stands out from the wall is z–5. The problems here are to produce the shape of the wall piece which is hyperbolic, and to which the tops of the rafters will be fixed; to develop the shapes of some of the rafters; and develop the surface of one half of the roof.

Having drawn the plan of the bay roof it is necessary to draw the plan and elevation of the cone of which the bay roof is a part. The edges of the elevation of the cone must be drawn at an angle equal to the pitch of the bay roof. In this case the pitch is 45°. Next draw the plan of the wall piece and divide the curve of the wall plate into a number of parts, say eight. Number these 1, 2, 3, etc. Draw lines from these points, which are on the base of the cone, up to the top

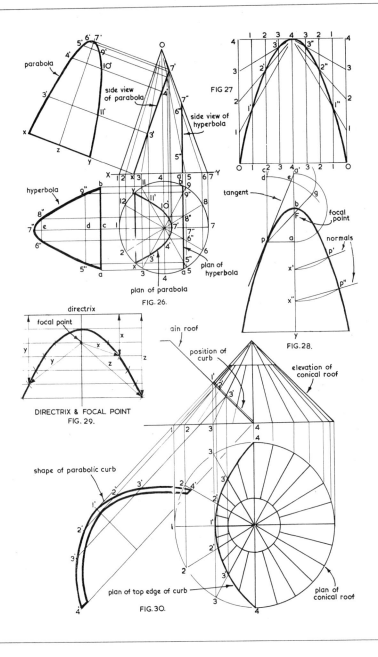

parabola

side view of parabola

side view of hyperbola

FIG 27

O

tangent

focal point

normals

FIG.28.

hyperbola

plan of hyperbola

plan of parabola

FIG. 26.

directrix

focal point

DIRECTRIX & FOCAL POINT
FIG. 29.

ain roof

position of curb

elevation of conical roof

shape of parabolic curb

plan of top edge of curb

plan of conical roof

FIG. 30.

point w. Project the points upwards vertically to the base of the cone in the elevation to give points 1, 2, 3, etc. From here they should be projected up to the top point of the cone, w, in the elevation.

Lines projected upwards from the plan from where the lines on the surface of the cone pass through the front edge of the wall piece, b, c, d, etc., to the appropriate line in the elevation, give the outline of the top edge of the wall piece. Only one point, e, will not be produced thus, and to obtain this point, place the compass point in w in the plan, and with radius w–e describe an arc to give e′ on the horizontal line from w. Project e′ upwards to give e″ on the edge of the elevation, and from here horizontally over to the centre line of the elevation to give point e.

The method of developing two of the roof rafters is fairly simple and should be well understood by the reader. Take rafter B. With compass point in w swing all the points on rafter B round to the horizontal line passing through w, and project these points upwards to where the rafter is to be developed. Make the pitch of the rafters equal to the slope of the elevation edge, in this case 45 .

To develop half the surface of the roof, open the compasses the length of one edge of the elevation and describe an arc. Step off along the arc a number of spaces equal to those half way round the plan of the roof and number these 1, 2, 3, etc. Join these points with straight lines to w.

The true distances from w down to b, c, d and e must now be marked on the appropriate line in the development. For instance, the true distance from w down to b is found by projecting a horizontal line from b in the elevation across to the edge of the cone to give point b′. w–b′ is the distance required. The distance from w down to e is w–e″. If all the points on the surface are treated thus, points b, c, d and e will be obtained on the development so that the drawings can be completed.

Conical turret. Fig. 32 shows the plan and elevation of the walls of a conical turret placed centrally on a roof pitched at 45 . The problems are to develop the shape of one curb on the main roof, and to develop the shape of the plywood required to cover the wall of the turret. These developments should be straightforward for the reader who has studied the foregoing drawings on the cone and should need little explanation. The curb is elliptical in shape, of course, and to produce

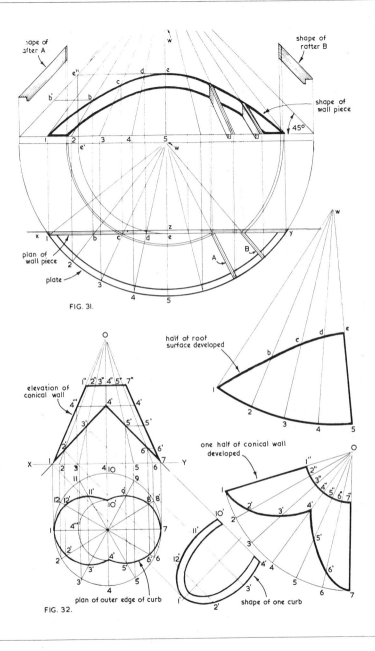

shape of
rafter A

shape of
rafter B

e''

shape of
wall piece

45°

b'   b

1   2   3   4   5   e'   w

x   1   b   c'   d   z   e   y

plan of
wall piece

plate

2

3

4

5

FIG. 31.

O

half of roof
surface developed

elevation of
conical wall

1" 2" 3" 4" 5" 7"

4"   4'   4'

3'   5'   5'

2'   6'   6'

X   1   7   Y

2   3   4 10   5   6

11

12 12'   11'

10'

1   4'''   7

2'

2   6' 6

3'   4'   5'

3   4   5

b   c   d   e

1

2

3   4   5

one half of conical wall
developed

O

1"

2"

3"   4" 5" 6" 7"

1

2   2'

3'

10'   4'

11'   5'

12'   4' 4   6'

3'   5   6   7

plan of outer edge of curb

shape of one curb

FIG. 32.

325

it one has first to draw the plan of the intersection of the turret wall with the main roof. This is seen in the plan. The development of half the surface of the wall is seen to the right of the drawings. Remember that to find points 2′, 3′, 4′, 5′, and 6′ in the development one has to project lines from 4′, 5′ and 6′ over to the edge of the elevation of the cone and measure down the edge from o.

Fig. 33 is the plan and elevation of a cone placed on the corner of a square pyramid. The problems are to complete the plan and elevation, and to develop the surface of the cone above the pyramid. First draw the plan and elevation as far as possible. Divide the plan of the cone into, say, twelve equal parts, and project these points up to its base in the elevation and from here to the point of the solid.

It will be seen that lines 1 and 4 go down to the edge of the pyramid, and line 2 comes down to point 2′. The position of this point can be seen in the elevation. This can be projected down to o–2 in the plan to obtain its position in that drawing. As o–3 line is the same distance from the corner of the pyramid as o–2 line, but on the other surface of the pyramid, the distance o–3′ is exactly the same as o–2′. Consequently a horizontal line from 2′ across to o–3 line will give the position of 3′ in the elevation. A vertical line downwards from 3′ into the plan will give its position in that drawing. All the points in the drawing can be obtained thus.

Only two points, namely x′ and y′, are not obtained thus. These can be positioned by assuming two more lines have been placed on the surface of the cone, o–x and o–y. Where o–x passes across the pyramid corner in the elevation will give point x′ and where o–y crosses the corner farther up will be point y′. These two points dropped down into the plan will give x′ and y′ in this drawing. To the right of the drawings is the development of the conical surface above the pyramid and is constructed as in other drawings.

Sphere. Another solid which crops up in carpentry is the sphere. Fig. 34 is the plan and elevation of a hemisphere, which of course is exactly half of a sphere. Two methods are used for approximately developing the surface of a sphere. Let us take that shown in the plan first. Draw the plan and elevation and divide the plan up into, say, six equal parts. To develop one of these sections, divide half of the elevation curve also into six equal parts and drop these points down and across the section to be developed. Project the edges of the section

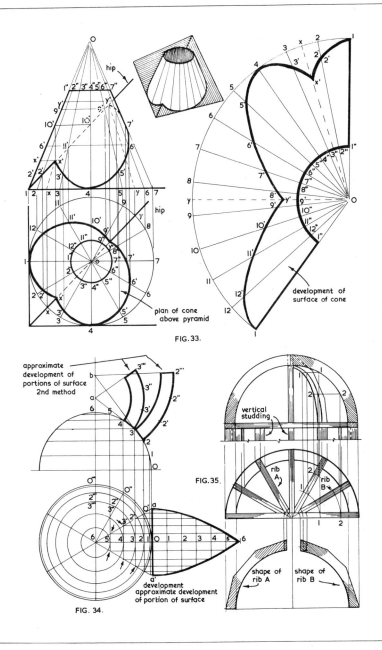

hip

plan of cone
above pyramid

FIG. 33.

development of
surface of cone

approximate
development of
portions of surface
2nd method

FIG. 35.

vertical
studding

rib
A

rib
B

development
approximate development
of portion of surface

FIG. 34.

shape of
rib A

shape of
rib B

to be developed out to intersect the line brought down from o to give points a and a'.

On the centre line of the development mark off the six spaces equal to those round the elevation, and draw vertical lines through these points to intersect with the horizontal lines brought out from the points on the edges of the section in the plan. The intersections thus obtained are points on the approximate development of one-sixth of the surface of the hemisphere.

The second method, which is also approximate, is to divide the surface of the hemisphere into horizontal strips. Each strip is considered to be a portion of the surface of a cone. To obtain the point of each cone, a line must be drawn through the two points at the extreme edge of the portion being developed and extended up to the centre line. For instance, let us consider the portion between points 2 and 3 in the elevation. Draw a line from 2 through 3 and onwards to meet the centre line in b. This is the top point of the imaginary cone and b–2 its edge. With compass point in b and radii b–3 and b–2 describe arcs, and mark off the distances 2–2', 2'–2'', 2''–2''' on the arc from 2 seen in the plan. Join 2''' to b. This is a development of the strip of surface one-fourth the distance round the hemisphere at that height.

Fig. 35 is a practical example based on the hemisphere, and shows a wooden niche frame. It shows how the elevation can be completed and how the intermediate ribs are developed.

Conical linings. Fig. 39 is another example of work based on the cone. It represents the plan and elevation of splayed linings with a semi-circular head. The head consists of two semi-circular pieces, one at the front and the other at the rear, each made in two halves and connected with handrail bolts. They are joined by four rails which run from the front to the back of the head, as seen in the elevation. At A and B in the elevation is given the method of obtaining the width of the pieces of material from which the rails are cut. Fig. 40 shows the two templets required for cutting out the rails. The dimension for these are obtained from the plan. Fig. 41 illustrates the application of the templets and the bevels, also obtained from the plan, enabling the rails to be shaped correctly.

To obtain the shapes of the thin plywood panels, the face of the panel in the plan should be extended to the centre line to give point

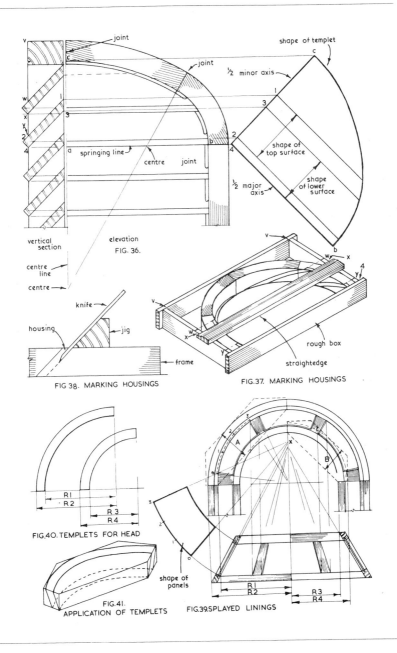

joint

joint

shape of templet

c

½ minor axis

1

3

2

shape of
top surface

shape of
lower
surface

½ major
axis

b

v

c

w

x
X
2
4

1

3

a springing line

centre joint

vertical
section

elevation
FIG. 36.

centre
line

centre

knife

housing

jig

frame

FIG 38. MARKING HOUSINGS

v

w x

y 4

w

x

y

rough box

straightedge

FIG.37. MARKING HOUSINGS

R1
R2

R 3
R4

FIG.40. TEMPLETS FOR HEAD

shape of
panels

FIG. 41.
APPLICATION OF TEMPLETS

3

2

1

A

x

B

o

o

3

2

1

o

R1
R2

R3
R4

FIG.39. SPLAYED LININGS

x, and using x for a centre, and opening the compasses to the front
edge and the back edge of the panel in turn, describe two arcs. On
that which starts from the front edge, mark off three spaces equal to
those seen in the elevation of the panel. Join point 3 to x to obtain
the shape of the three panels.

Louvre boards. Fig. 36 shows a method of obtaining the shape of the
templet required for setting out the louvre boards to an elliptical-
headed frame, and should be consulted when reading the text for the
circular louvre frame Fig. 44.

Figs. 42, 43 and 44 show three frames for ventilation purposes.
The first, Fig. 42, is a triangular frame with louvre boards, and the
problems are to develop the shape of the templet from which all the
boards can be obtained, and also the positions of the housings on the
two inclined sides. Let us take the latter first.

Draw the elevation and a section through the frame, placing in the
section the positions of the boards. Now take the top board A. The
housing for this board is from point 1 to e at the top edge and from
3 to 4 at its bottom edge. These four points should be projected over
to the inside surface of one of the sides in the elevation, and from
there over to the edges of the developed surface seen just above the
side of the elevation. This gives points 1, e, 3 and 4 on this drawing.
Join 1 to e and 3 to 4 with straight lines to obtain the correct position
of the housing for board A.

Another example is the louvre board C. Project points w, x, y and z
over to the inside surface of the side in the elevation, and from these
points over to the appropriate edge of the developed side to obtain
the position of the housing.

To develop the shape of the templet, which in this case is triangular,
project points a and b on the centre line of the elevation over to where
it is required to develop the templet. Draw the centre line of the
templet, a′–b′, at the same inclination as the boards in the frame, and
make the bottom edge of the templet at right angles to its centre line
and equal in length to c–d in the elevation. Join a′ to c and a′ to d.
This is the shape of the templet, and should be cut from a piece of
plywood or hardboard.

The shapes of the top and bottom surfaces can now be marked on
the templet. To obtain the shapes of the surfaces of board A, project
points 1, 2, 3 and 4 over to the centre line of the templet, and from

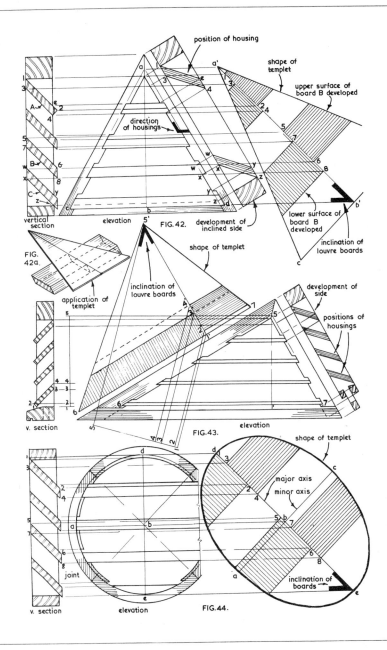

position of housing

shape of templet

upper surface of board B developed

direction of housings

development of inclined side

vertical section · elevation · FIG. 42.

lower surface of board B developed

inclination of louvre boards

FIG. 42a.

application of templet

shape of templet

inclination of louvre boards

development of side

positions of housings

v. section · elevation · FIG. 43.

shape of templet

major axis

minor axis

joint

inclination of boards

v. section · elevation · FIG. 44.

these points over to the edges of the templet at right angles to the centre line. The distance between lines 1 and 2 is the shape of the top surface of the board, and the distance between points 3 and 4 is the shape of the lower surface. The shapes of the board B are also seen on the templet.

Fig. 42a shows how the templet is applied to the board material. Suppose board B is to be marked out. First obtain a piece of timber long enough for the purpose, and apply the bevels to the front and rear edges. Square a centre line round all the four surfaces of the board. With the front surface of the board facing upwards, place the templet on the board, making sure that the centre line of the templet is directly over the centre line of the board. Also ensure that the lines 5 and 6 on the templet are directly over the top and lower edges of the board. The ends of the board can now be marked along the edges of the templet directly over the board.

Now turn the board over so that the back surface faces upwards. Place the centre line of the templet directly over the centre line of the board, but this time make sure that the lines 7 and 8 are directly over the top and bottom edges of the board. The ends of the board are now marked again and the templet removed. The ends of the board are cut to the lines on the top and lower surfaces and then fitted into the frame. Treat the other boards in the same way, making sure that the correct lines on the templet are used for each board.

The second triangular louvre frame, Fig. 43, is dealt with in exactly the same way. The templet in this drawing has been placed in the position shown because of lack of space. Normally, the line 1–5, below the templet, would be in the same position as line 1–5 just to the right of the vertical section.

Although the circular louvre frame, Fig. 44 is an entirely different shape from the last two, the development of the templet is almost the same. Draw the elevation and the section through the frame and project points d and e over to where the templet is to be drawn. The centre line of the templet, d–e, is drawn at the same angle as the louvre boards and the other centre line a–c is equal to the width a–c in the elevation. The shape of the templet is elliptical and the board shapes are placed on the templet in the same way as for the triangular louvre frames.

The drawings in Fig. 36 should be fairly straightforward because the geometry for this semi-elliptical headed louvre frame is the same as for the circular frame. Only half of the required templet is shown. To mark the housings on the curved inside surface of these frames take the frame in Fig. 36 as an example. A rectangular box should be made, Fig. 37, so that the curved head just fits into it. On the sides of the box from point v should be marked the distances seen in the section, Fig. 36, i.e. v–w, v–x, v–y, etc. A straight edge placed across each pair of points, w–w, x–x, etc. enables the positions and direction of the housings to be marked on the inside surface of the frame, see Fig. 37. The housings are marked by placing a jig, Fig. 38, across each pair of lines, and marking the slope of the housings with a piece of steel sharpened to a knife edge. If the jig bevel is the same as the inclination of the louvre boards, and the knife is kept flat on the jig, as in Fig. 38, the housings marked on the inside surface of the frame will be at the correct angle.

Raking mouldings. When two horizontal mouldings intersect at a corner in a building the sections of the mouldings are exactly the same, but if one or both of the mouldings were inclined, their sections would have to be different.

Take the simple case shown in Fig. 45. The plan shows three mouldings, A, B and C. Let us assume that moulding A is horizontal, B is inclined at 30, and C is horizontal. To develop the shapes of B and C when A is known, draw a section of A immediately above its plan as shown, and from all the points on the moulding draw lines upwards at the inclination of moulding B. Draw o–o′ the back edge of moulding B, at right angles to the inclined lines representing the elevation of the moulding. Set off along the top line o–1–2–3 equal to the thickness of mould A, and drop lines across the elevation parallel to the back edge o–o′. The various intersections will give the shape of moulding B. If the inclined lines from moulding A are carried upwards to where moulding C is to be developed, the intersections will again be obtained by making the distances across the moulding equal to those on mouldings A and B, and dropping lines vertically to meet the inclined lines.

Mitre bevels. To develop the mitre bevels, take moulding A first. To obtain the bevel so that it mitres with B correctly, its top surface must be developed. With compass point in o and radius o–x describe an arc to give x′ on the horizontal line o–x′. Drop a line down to the plan to meet a horizontal line brought out from x to give x′

in this drawing. The bevel required to apply to the end of A is indicated at x'.

To obtain the bevels to apply to the ends of moulding B, its top surface must be developed. This is shown in the elevation. With compass point in o on the section of mould B, and with radius o–y describe an arc to give point y' on the extended back edge of the section. Draw a line through y' parallel to the inclined moulding to meet lines brought up at right angles to the inclination to give points x'' and z. The bevels are also indicated. The bevel to apply to the end of mould C is seen in the plan because the top surface of the moulding is in the horizontal plane.

Fig. 46 shows the plan and elevation of two mouldings intersecting at an angle of 135 . Mould A is horizontal and B is inclined at 30 . The shape of mould A is known. Draw the plan of the two mouldings and the elevation of the horizontal moulding A. Note that when an inclined moulding is involved with a horizontal moulding, the inclined moulding should be drawn parallel to the top and bottom edges of the drawing sheet. Draw a section of mould A on the elevation, and mark the various distances o–1–2–3 across its thickness on the plan of the moulding. Project these points on the plan over to the mitre and upwards vertically to meet the horizontal lines brought over from the section in the elevation, so that an elevation of the mitre can be constructed. From all the points on the mitre elevation draw lines parallel to the inclination of moulding B and draw the back edge of this, o–o', at right angles to the inclined lines.

Make the distances across moulding B equal to those across A, o–1–2–3, and drop lines from these points parallel to the back edge to obtain a series of intersections. Join these up as shown to obtain the shape of moulding B.

Fig. 48 shows the intersection of two mouldings, A and B, both inclined at 30 , the shape of moulding A being known. Draw the plan of the two mouldings and the elevation of mould A. Note that when two inclined mouldings are involved, the plan of the moulding of known shape is drawn parallel to the top and bottom edges of the drawing sheet. Draw the section of the known moulding and mark off the distances seen in the section across its plan. Project the points across the plan over to the mitre line, and upwards vertically to meet the inclined lines brought up from the elevation section. These intersections will give an elevation of the mitre.

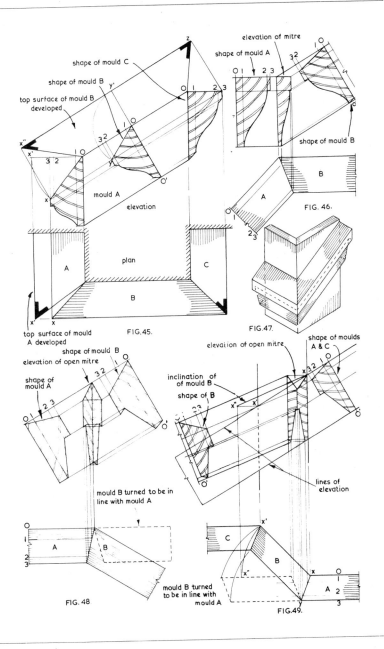

elevation of mitre

shape of mould A

shape of mould C

shape of mould B

top surface of mould B developed

z

y'

O 1 2 3

3 2

x"

x'

3 2

1

O

x

O'

mould A

elevation

3 2 1

O

shape of mould B

B

A

2 3

O

FIG. 46.

plan

A

B

C

x' x

top surface of mould A developed

FIG.45.

FIG.47.

elevation of open mitre

shape of mould B

shape of mould A

O 1

3 2

2 3

O'

mould B turned to be in line with mould A

O

1

2

3

A

B

FIG. 48

mould B turned to be in line with mould A

elevation of open mitre

shape of moulds A & C

inclination of of mould B

shape of B

3 2 1

x 3 2 1

O

x"   x'

O'

lines of elevation

x'

C

B

x"

x

O
1
2
3

A

FIG.49.

335

life of the carpenter and joiner. Roof work, hand-railing, and other forms of work often call on the development of oblique planes for the correct setting out of such work. Fig. 52 illustrates what an oblique plane looks like, and Figs. 50 and 51 show how to find the inclination of such a plane to the horizontal and vertical planes.

Let V.T and H.T, Fig. 50, be the vertical and horizontal traces of an oblique plane. To find its inclination to the horizontal plane it is necessary to develop the shape of a right-angled triangle placed beneath the plane so that its base a–b is at right angles to the H.T. and its hypotenuse b–c, is in contact with the surface of the plane. To develop the triangle let a–b be the base of the triangle a–b–c. a–b can be placed anywhere along the H.T. and at right angles to it. From a draw a vertical line to meet the V.T. in c. This is the second side of the triangle. With compass point in a and radius a–b describe an arc to give b′ on the x–y line. Join b′ to c. b′–c is the hypotenuse of the triangle and a–b′–c its inclination to the horizontal plane.

To obtain the inclination of an oblique plane to the vertical plane, Fig. 51, a right-angled triangle must again be developed, this time with its base at right angles to the x–y line but inclined over so that the side a–c is at right angles to the V.T. This will again make the hypotenuse b–c in contact with the oblique surface. To develop the triangle place the side a–b anywhere so long as it is 90° to the x–y line. From a, and at right angles to the V.T., draw side a–c. With a as centre and radius a–c describe an arc to give c′ on the x–y line. Join c′ to b. The developed triangle is a–b–c′, and angle b–c′–a is the inclination of the oblique plane to the V.P.

In Fig. 53, a–b–c–d is the plan and elevation of an oblique plane, and Fig. 56 a pictorial view of the surface. The problems are to develop the shape of the surface and its inclinations to the V.P. and the H.P. To develop the shape of the surface draw the plan and elevation. The true length of sides a–b and c–d can be seen in the elevation because they are parallel to the V.P., but the lengths of sides a–d and b–c must be developed. If we develop the length of a–d this will also give the length of b–c because they are equal. Place the compass point in d in the plan, and with radius a–d describe an arc to give a′ on the x–y line. Join a′ to d in the elevation to obtain the lengths of a–d and b–c.

We now require to develop the distance between two opposite corners, say a and c. To find the length of diagonal a–c place the

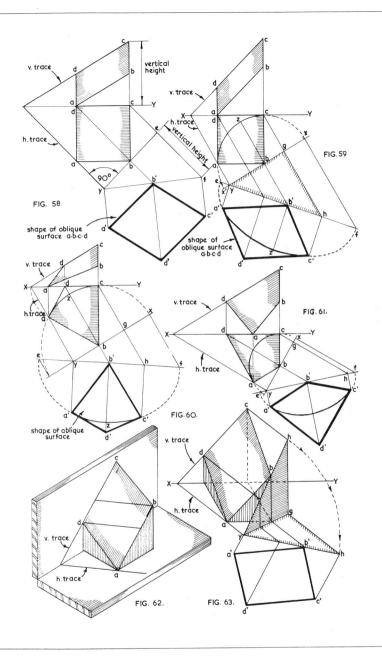

FIG. 58

vertical height

v. trace

h.trace

90°

shape of oblique
surface a·b·c·d

FIG. 59

v. trace

h.trace

vertical height

shape of
oblique surface
a·b·c·d

FIG. 60.

v. trace

h.trace

shape of oblique
surface

FIG. 61.

v. trace

h. trace

FIG. 62.

v. trace

h. trace

FIG. 63.

v. trace

h. trace

Moulding B must now be turned so that it is in line with moulding A, see plan. If this is done the mitred surface of the two mouldings will be seen in the elevation, and the mitred surface of B is exactly the reverse of A.

Draw the second half of the open mitre, and project lines from all the points of mitre B at the angle of its inclination. Draw o–o′, the back edge of moulding B, at right angles to the inclined lines, and make the distances o–1–2–3 equal to those on moulding A. Drop lines from these points to intersect with the inclined lines to obtain the shape of moulding B.

Fig. 47 shows a pictorial view of two mouldings, one inclined and the other horizontal, and clearly shows that if the two mouldings are to intersect correctly, they must have different sections.

Fig. 49 shows three mouldings, A, B and C, intersecting as in the plan. The line of the elevation indicates that the three mouldings must appear in a straight line when viewed from the front. A problem such as this could occur on the wall adjoining a staircase where the mouldings follow the pitch of the stairs. Mouldings A and B could be mitred round a projection on the wall surface.

As already mentioned, the mouldings must appear to be in a straight line, but it should also be realized that mould B is inclined at a different angle to the other two. The inclination of mould B must, therefore, be developed. Assume that the shape of mould C is known. As moulding A runs parallel to moulding C, the shape of moulding A is the same as C. Draw the plan to any given shape; also the lines of the elevation. Add the section of A between the elevation lines and above its plan. Draw the open mitre of moulds A and B as shown in Fig. 48.

To develop the inclination of mould B, place the compass point in x, in the plan, and with radius x–x′ describe an arc to give x″ on the horizontal line brought out from x. Project a line upwards from x″ to intersect with a horizontal line brought out from x′ in the elevation to give x″ in the drawing. Join x″ to x. This line represents the inclination of moulding B. Draw lines from all the points on the open mitre of B parallel to its inclination line x–x″, and develop its shape between these lines as described in the other drawings and as indicated to the left of the elevation.

Oblique planes. These and their development play a big part in the

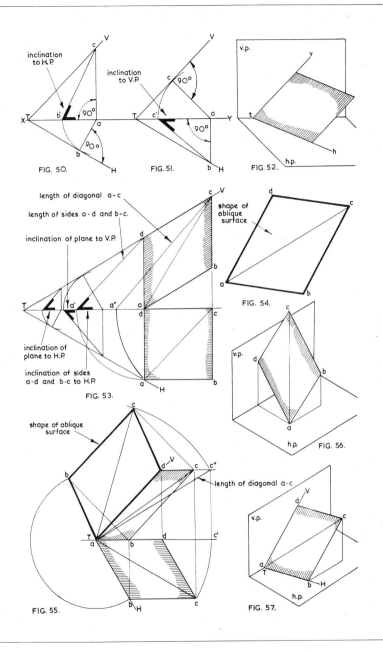

inclination to H.P.

90°

90°

FIG. 50.

inclination to V.P.

90°

90°

FIG. 51.

v.p.

h.p.

FIG. 52.

length of diagonal a-c

length of sides a-d and b-c.

inclination of plane to V.P.

inclination of plane to H.P.

inclination of sides a-d and b-c to H.P.

FIG. 53.

shape of oblique surface

FIG. 54.

v.p.

h.p.   FIG. 56.

shape of oblique surface

length of diagonal a-c

FIG. 55.

v.p.

h.p.

FIG. 57.

339

compass point in c in the plan and with radius c–a describe an arc to give a'' on the x–y line. Join a'' to c to give the length of the diagonal. To draw the shape of the surface, Fig. 54, first draw the line a–c. With compass point in a and open the length of a–b describe an arc in the direction of b. With the same radius and compass point in c make an arc in the direction of d.

Open the compass the developed length of a–d and b–c, and with the point in a make another arc to cut the arc at the top, giving point d. With compass point in c cut the other arc to give the position of b. a–b–c–d, Fig. 54, is the developed shape of the surface a–b–c–d, Fig. 53.

To develop the inclinations of the surface to the V.P. and the H.P. the vertical and horizontal traces must first be drawn. Continue the line c–d, Fig. 53, downwards until it meets the x–y line to obtain the V.T., and from T draw a line through a to give the H.T. To obtain the inclinations follow the instructions given in Figs. 50 and 51.

Fig. 55 shows another way of developing the shape of an oblique plane a–b–c–d. The shape is shown at the top of the drawings, and the surface has been swung round on edge a–d into the upright position so that its true shape can be seen. First develop the length of the diagonal, as in Fig. 53, then with compass point in a and radius a–c'' describe an arc in the direction of c in the development. From c in the elevation draw a line at right angles to a–d (the hinge line) to give the position of c in the development.

Draw a line from b in the elevation parallel to c–c to meet a line from c in the development parallel to a–d to complete the shape of the surface. As a–d is in contact with the V.P. this line is also the V.T. of the surface. Similarly, as a–b is in contact with the H.P. this is also the H.T. of the surface. Fig. 57 is a pictorial view of the surface.

Oblique planes for handrailing. Now we come to the method of developing oblique planes used in handrailing. Fig. 58 is the plan and elevation of an oblique plane a–b–c–d, similar to that shown in Fig. 56. To develop the surface draw the V.T. and the horizontal trace, also another line through b and at right angles to the H.T. Draw lines from b, c and d parallel to the H.T., and on the line from c, and, measuring from e, mark off the vertical height of the surface. This is taken from the elevation. Join f to y and from y, b' and f draw lines at right angles to y–f, making y–a' equal y–a, b'–d' equal

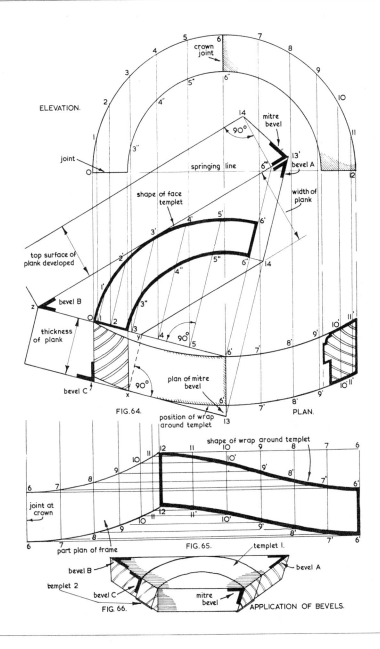

ELEVATION.

crown joint

joint

O

springing line

mitre bevel

90°

14

13'

bevel A

width of plank

shape of face templet

top surface of plank developed

bevel B

z

thickness of plank

bevel C

x

90°

90°

plan of mitre bevel

position of wrap around templet

13

FIG. 64.

PLAN.

shape of wrap around templet

joint at crown

part plan of frame

FIG. 65.

templet I.

bevel B

templet 2

bevel C

FIG. 66.

mitre bevel

bevel A

APPLICATION OF BEVELS.

to b–d, and f–c′ equal to e–c. a′–b′–c′–d′ is the shape of the oblique surface.

Fig. 59 shows another and similar oblique plane developed, and when this is compared with the geometry shown in Fig. 58, it will be seen to be the same in almost every detail. Now study Fig. 63, where this geometry of the oblique plane is illustrated. In addition to the surface being developed in Fig. 59, the curve a–c has also been placed on the development. This is particularly useful because, in hand-railing, the templets for applying to the timber are developed in this way.

It should be remembered that any line placed across the surface and parallel to the horizontal trace is a level line, and can be measured in the plan. Place on the drawing the curve a–c and extend it round to x and x′. From these points project it down to the y–h line to give e and f. The distance e–f is the major axis and b–z, seen in the elevation, is the minor axis of an ellipse which will give the shape of the curve a–c on the developed surface.

From b′ on the development make b′–z equal b–z on the elevation and, using these points, construct a semi-ellipse to pass through points e, a′, z, c′ and f.

Fig. 60 is the plan of a plane in the shape of a trapezium. If the reader compares this drawing and Fig. 61 with the drawings in Figs. 58 and 59 he will find that the geometry for these two drawings is exactly the same.

Fig. 62 is another pictorial view of an oblique plane, and shows that any line on the surface of the plane, so long as it is parallel to the H.T., is a level line and its true length can be seen in the plan.

Double curvature. Finally we come to the subject of double curvature or, as some people describe it, circle on circle work. The first example involves a frame which is curved in plan and has a semi-circular head in elevation. The jambs of the frame are parallel. The second example is similar to the first except that it has radiating jambs.

Fig. 64 is the plan and elevation of a frame in double curvature with parallel jambs. The problems are to decide on the minimum thickness and width of the material required, and the shape of the templets necessary for shaping the head, and to develop any necessary bevels. Draw the plan and elevation to the dimensions of the frame and divide the elevation curve into any number of equal parts, say twelve. Draw a line to connect points O and 6′ in the plan, and another line parallel to this to just touch the outside edge of that half of the frame. The distance between these two lines represents the minimum thickness of the timber required to produce the two halves of the head of the frame.

Project vertical lines down from the points round one half of the elevation, to intersect with the o–6′ line in the plan. Draw lines from all these points at right angles to o–6′, and make the lengths of these lines equal to those in the elevation, measuring from the o–12 or springing line. Mark off, too, the heights of the 3″, 4″, 5″ and 6″ points on these lines. Freehand curves through these two sets of points will give the shape of the face templet required for the work.

The minimum width of the material is found as follows. Draw a line from x in the plan at 90° to the edges of the plank to give point y on the opposite side. Next draw a line from 13 in the plan parallel to 6′–6″ to meet a line parallel to o–6′ brought out from 6″, giving point 14. Also put in a line from y to 14 to give one edge of the plank. Draw another line parallel to y–14 to just touch the outside edge of the templet shape. The distance between these two lines is the minimum width of timber required for the head.

The mitre bevel is found by developing the shape of the top surface of the plank, and this is shown above the development of the templet.

When the setting out has been completed two pieces of timber will be required, one left and one right hand, shaped to y–14–13′–z, seen around the templet shape, each piece being equal in thickness as that seen in the plan. With the mitre bevel applied to the top end of each, the two halves can have the face templets applied to the front and back surfaces. Fig. 66 shows how these and the bevels are applied, and the two halves shaped to the outline of the templets. When this has been done the wrap-around templet can be applied round the outside edge of each half to obtain their shapes seen in the plan.

To develop the shape of the wrap-round templet, Fig. 65, project the points 6 to 12 around the edge of the elevation downwards to give points 6 and 6, 7 and 7, 8 and 8, etc. on the front and rear edges of the frame in plan. Project horizontal lines from all these points over towards the right and on the line from 12 step off six distances equal to those round half the elevation. Drop lines downwards from the points obtained to intersect with the horizontal lines brought over from the plan to give 11′ and 11′, 10′ and 10′, 9′ and 9′, etc. Freehand curves through these two sets of points will give the shape of the wrap-around templet.

The templet should be made from some thin, flexible material such as $\frac{1}{8}$ in. plywood. When placed correctly it should give the outline of the curve as seen in the left hand half of the plan, Fig. 64. Each half of the frame can then be shaped, and checking that the surface is always vertical when the piece being shaped is stood up on its lower joint surface.

Cuneoid problems. Fig. 67 is the plan and elevation of a semi-circular-headed frame, curved in plan and with radiating jambs. Only a brief outline of the geometry involved can be dealt with in this chapter. This frame head is based on the geometrical solid known as the cuneoid, see Fig. 70. The outline of the cuneoid in the plan, Fig. 67, is o–6–a–z. The elevation of the front face of the solid is the semicircle o–6–a, and its back edge is h–n.

To develop the bevels, templets, etc. draw the plan and elevation and divide one half of that drawing into, say, six equal parts. Draw a line across the o''–6'' points in the plan, and then another parallel to it to just touch the outside edge of the plan of the frame. The distance between these two lines is the minimum thickness of the timber required.

Drop lines from the points round the outside of the elevation down to the o–6 line (this is the front face of the cuneoid in the plan) to give points 1, 2, 3, etc. on this line. From z, the back edge of the cuneoid, draw lines through o, 1, 2, 3, etc. to give points o'', 1'', 2'', 3'', etc. on each edge of the plank in the plan. From these points draw lines at right angles to the plank edges, and make these lines equal to those in the elevation. The broken line in the elevation represents the inside edge of the frame and these various heights can also be placed on the appropriate lines in the templet development. Freehand curves through the points obtained will give the shapes of the two templets required for this type of frame. The outline of the plank which gives the minimum width, and the development of the mitre bevel, are made in the same way as for Fig. 64, but for accuracy the development of the plank width should be made on the convex side of the plan. To obtain point y draw a line from x at right angles to the plank edges to give y on the convex side of the frame.

Fig. 69a shows how the two templets are applied to each half of

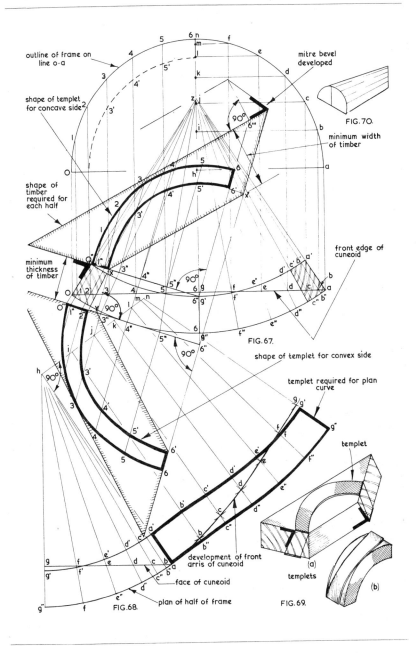

outline of frame on line o·a

shape of templet for concave side

shape of timber required for each half

minimum thickness of timber

mitre bevel developed

FIG. 70.

minimum width of timber

front edge of cuneoid

90°

90°

90°

90°

90°

90°

FIG. 67.

shape of templet for convex side

templet required for plan curve

templet

FIG. 69.

(a)

templets

(b)

development of front arris of cuneoid

face of cuneoid

plan of half of frame

FIG.68.

the head, and at b is shown the wrap-around templet applied to one of the partly-shaped halves.

Wrap-around templet. We now come to the setting out of the wrap-around templet, and this first involves developing the surface of the cuneoid, Fig. 68. This drawing, which is really the right hand half of the plan, Fig. 67, has been drawn separately to avoid confusion. Briefly, one should follow these steps. Draw vertical lines downwards from points a–b–c, etc. on the elevation to give points a, b, c, etc., on the front edge of the cuneoid in the plan. We should now show the position of this front arris or corner in the development. Draw a line at 90 to a–h from h in the plan and mark off heights h, i, j, k, etc., equal to those in the elevation. This is the back edge of the cuneoid.

To obtain point b on the arris, place the compass point in a and with radius a–b, taken from the elevation, make an arc in the direction of b in the development of the arris. Then with compass point in i (this is on the back edge of the cuneoid) and with radius h–b in the plan, Fig. 68, make an arc to intersect with the first to give the position of point b. To obtain the position of point c on the arris of the cuneoid, with compass point in b (on the arris) and radius b–c (in the elevation) make an arc in the direction of c. With compass point in j (back edge of the cuneoid) and radius h–c (on the plan) cut the other arc to give the position of c on the arris. All the points on the arris can be obtained thus, and they must all be obtained before the shape of the templet can be developed.

Having obtained all the points on the arris the templet shape can be arrived at. a and a′ are two points at the end of the templet. To obtain b′ and b″ place the compass point in b on the arris, and with radii b–b′ and b–b″ taken from the plan, mark off these distances on line b–i to obtain the two points required.

To obtain points c′ and c″ on the templet, with compass point in c on the arris, and with radii c–c′ and c–c″ taken from the plan, mark off these distances on line c–j to obtain the two required points. Repeat until all the points on the templet development have been plotted. Two freehand curves through these sets of points will give the shape of the wrap-around templet. It is applied as in Fig. 69b.

Index

A

Acute angle, 59
African mahogany, 22
Air drying, 15
Angle brackets, 135
  joints, 42
Applied geometry, 308
  Archimedean spiral, 310
  cone, 321
  cuneoid, 344
  cylinders, 318
  double curvature, 342
  drop scroll, 310
  linings, 329
  louvres, 330
  oblique planes, 336
  parabola, 321
  prisms, 313
  pyramids, 314
  raking mouldings, 332
  scroll, 310
  solid geometry, 313
  sphere, 326
  tangents, 308
Apron moulding, 139
Arcade, 191
Arches, 94
Architraves, 132
Astragal, 66

B

Back iron, 27
Baltic whitewood, 21
Bark, 9
Bast, 9
Beam box, 91
Beetles, 9, 12
Bench hook, 38
Bevel-edged chisel, 33
Bevelled shoulder, 30
Binders, 120

Bird's mouth, 119
Bisection, 59
Block plane, 28
Bolection moulding, 152
Bolts, 58
Bowing, 20
Bow saw, 26
Braces, 32, 149, 158
Bracing, 100
Bradawl, 32
Bridging joint, 108
British Columbian pine, 21
Broad-leafed trees, 7
Bullnosed plane, 28
  step, 173
Bullseye opening, 98
Butt joint, 39
Button, 43

C

Cambium cells, 9
Capillary grooves, 139
Case hardening, 18
Cavetto, 66
Ceiling joist, 113, 120
Centre bit, 32
Centres, 94
Centres, double curvature, 201
  semi-circular, 199
  semi-elliptical, 199
Chamfer, 30, 152
Circle, 63
Circumference, 63
Claw hammer, 36
Cleat, 183
Closed eaves, 123
Clout nail, 49
Coach screw, 51
Collar roof, 116
Collars, 120
Column box, 93
Common rafter, 119

Compasses, 36
Compass saw, 27
Concrete formwork, 207
  formwork canopy, 212
  formwork stairs, 208
  formwork wall, 213
Cone, 73
Cone bearing trees, 7
Connectors, shear plate, 265
Connector, toothed, 265
  split ring, 265
Conversion, 12
Coping saw, 27
Co-pitch, 237
Corner joints, 39
Counter cramp, 45
Counters, 274
Countersink bit, 32
Countersunk screw, 51
Couple close roof, 116
Cover moulding, 135
Cramps, 93, 136
Crosscut saw, 23
Cross halving, 120
Cuneoid, 344
Cupping, 13
Cup shake, 20
Cutting list, 78
Cylinder, 73
Cylinder lock, 32

  panelled, 242
  vestibule screen, 240
  with fanlight, 244
Door frames, 56, 149
  linings, 56, 135
Double curvature, 342
Double floors, 110
Douglas fir, 21
Dovetails, 45
Dowel bit, 32
Draw boring, 173
Drying schedules, 18
Dry rot, 20, 100
Durium-tipped drill, 52

E

Eaves, 119, 123
Ellipses, 65
Elliptical arch, 94
Elm, 22
End grain, 19
  shakes, 20
English oak, 22
Equilateral arch, 94
  triangle, 60
European redwood, 21
Expansion bit, 32
External frame, 161

D

Damp-proof course, 15, 20, 100
Dead shoring, 185
Decay, 20
Deciduous, 7
Defects, 19
Degrades, 15, 16
Diameter, 62
Diffuse porous, 11
Dihedral, 166
Dividers, 36
Doors,
  church, 246
  double margin, 240
  entrance, 157
  external, 244
  fire resisting, 246
  flush, 155
  glazed, 155
  ledged & braced, 149

F

Face marks, 78
Fanlight, 135, 136, 141
Fascia, 114, 123
  board, 162
Felling, 12
Fender wall, 100
Fibre plugs, 82
Fibres, 11
Firmer chisel, 33
Firring pieces, 131
Fish plates, 45
Fixing devices, 347
Floor boards, 12
  brads, 77
  joists, 12, 19, 100
  slabs, 91
Flooring joints, 39
Floors, 100, 105
Focal points, 66

Folding partitions, 252
Formwork, 85
Fungi, 9
Fungus, 12

G

Gable, 119
Gantries, 204
Glazing beads, 135
Glue blocks, 146
Glulam work, 300
  adhesives, 306
  assembling, 304
  cramping bed, 304
  cramps, 303
Going, 168
Gouges, 33
Gravity toggle, 83
Grinding angle, 28
Grindstone, 28
Ground floors, 100
Grounds, 142
Growth rings, 9, 12
Gullet, 26
Gusset plates, 199
Gutter board, 114

H

Half-round step, 174
Half-space landing, 174
Halving joints, 40
Hammer-headed key, 45
Handrail, 32, 171, 174, 267
  bevels, 294
  bolt, 45, 139
  marking out, 290
  one bevel work, 287
  placing templets, 294
  two bevel work, 292
Hardboard nail, 51
Hardwoods, 7
Haunching, 42, 151
Heart shakes, 20
Heartwood, 9
Heptagon, 63
Herring-bone strutting, 108, 131
Hexagon, 63
Hip rafter, 119
Honduras mahogany, 22

Horizontal section, 74
Housed joints, 40

I

Inclined plane, 162
In situ, 85
Insulating board, 132
Ionic volute, 311
Isometric, 69
Isosceles triangle, 60

J

Jack plane, 27, 28
  rafters, 119

K

Keyhole saw, 27
Kiln seasoning, 15
Kneww bend, 179
Knots, 19

L

Lagging, 97, 199
Lantern, light, 265
Lean-to-roof, 114
Ledges, 149
Length joints, 42
Letter plate, 157
Linings, 135, 142, 158
Lintel, 85, 88
Lippings, 155
Lost-head nails, 49
Louvres, 330

M

Major axis, 65
Mallet, 36
Marking gauge, 35
  out, 78
Matching, 151
Metal dogs, 187
Metal dowel, 56
  plug, 52, 56
Minor axis, 65

Mitre block, 38
  joint, 40
  square, 33
  templet, 38
Moisture content, 15, 17
Morse drill, 32
Mortice and tenon joints, 40, 151
  chisel, 33
  gauge, 35
Mouldings, 66
Mullions, 139

N

Nail punch, 38
Nails, 49
Needle, 183, 185, 188
Newel, 171
Nogging pieces, 132
Normal, 65, 66
Nosing, 171

O

Oblique mortice and tenon, 42
Obtuse angle, 59
Octagon, 63
Ogee roof, 268
Oilstone, 28, 38
Oregon pine, 21
Outer string, 171

P

Pacific Hemlock, 21
Panel pin, 49
  saw, 23
Panelling, dado, 269
  frieze, 271
Parabola, 321
Parabolic arch, 97
Parenchyma, 11
Paring chisel, 33
Parting bead, 145
Partitions, 132
Pentagon, 61
Piling sticks, 15
Pin bit, 32
Pitch, 126
Pith, 9

Pivot hung sash, 141
Pivots, 142
Planted mouldings, 152
Plates, 101
Plough, 30
Plywood panels, 135, 152
Plywood, resin bonded, 207
Pocket piece, 145
Pointed arch, 246
Poling boards, 196
Polygons, 60
Pre-cast, 85
Purlins, 119, 120

Q

Quadrant, 63
Quadrilaterals, 60
Quarter sawing, 12, 13
Quarter-space landing, 177

R

Radius rod, 98
Rafters, 114
Raised-head screws, 51
Rakes, 180
Raking mouldings, 332
Raking shores, 180
Ramp, 178
Rawlanchors, 55
Rawlbolts, 56
Rawlnuts, 55
Rays, 11, 13
Rebate plane, 28
Rectangle, 60
Red deal, 21
Redwood, 21
Resin bonded plywood, 207
Rhomboid, 60
Rhombus, 60
Ribs, 97
Ridge, 116, 119
Right angle, 59
Right-angled triangle, 60
Ring porous, 11
Rip saw, 23
Rise, 168
Riser, 167
Riser boards, 210
Rods, 74

Roof geometry, 125, 225
Roof lights, 257
Roofs, 114
  bowstring, 218
  flat, 116, 128
  hammerbeam, 222
  hipped, 119
  knee-braced, 218
  lean-to, 114
  low pitched, 214
  octagonal, 316
  ogee, 268
  open type, 220
  shed, 216
Round-head screws, 51
Router, 28

S

Sap, 7, 8
Sapwood, 7
Sashes, 142
Sawing stool, 38
Saws, 23
Saw set, 26
  sharpening, 24
Saw vice, 26
Scalene triangle, 60
Scales, 59
Scotia, 66
Scots pine, 21
Scraper, 38
Screw driver, 36
  bits, 32
Screws, 51
Scroll, 310
Seasoning, 13
Secret nailing, 7, 39
Sector. 63
Segment, 63
Segmental arch, 94
Semi-circle, 63
Semi-circular arch, 94
Setting out rods, 74
Shakes, 12, 20
Shoring, 180
  arched,187
  erection, 188
  flying, 188
  horizontal, 191
  raking, 187
  to column, 191

Shoulder plane, 20
Shrinkage, 12, 13, 18
Shutters, 93
Sill, 136
Single roofs, 114
Skirtings, 58, 135
Sleeper wall, 100
Sliding bevel, 33
  sashes, 142
Sliding door cupboard, 250
  garage, 250
  partitions, 252
Sliding door gear, 249
Slot-screwed joint, 39
Smoothing plane, 28
Soffit, 123
Softwoods, 7, 12
Solid geometry, 313
Span, 12, 94, 100
Splayed work, 162
Spokeshave, 30
Springing line, 199
Spring toggle, 55
Springwood, 9
Sprocket piece, 123
Square, 60
Squaring rod, 158
Stack, 15
Stairs, baluster rails, 171
  commode step, 284
  concrete formwork, 208
  curtail step, 283
  curved strings, 278
  dancing steps, 296
  dog-legged stairs, 173
  geometrical, 276
  half-round step, 174
  half-space landings, 174
  handrail, 32, 171, 174
  open newel, 177
  spiral, 286
  staircase well, 177
  staircase works, 168

Star shake, 20
Steel square, 288
Stickers, 15
Stile, 151
Stockings, 58
Storey rod, 168
Straight flight, 171
Straining piece, 191
Struts, 98, 120
Stuck moulding, 152

Stud partition, 132
Studs, 135
Summerwood, 9, 11
Swan neck, 178
Swelling, 13

T

Tangential, 12
Tangents, 62, 66, 308
T.D.A. roof, 117, 125
Tempered hardboard, 208
Templets, 97, 127, 173
Tenon saw, 23
Tension, 19
Through and through, 12
Tie, 98
Timber connector, 51, 119
Timbering, 196
Toggles, 52
Tongue and groove joint, 40
Torus, 66
Tracery, 66
Tracheids, 11
Transom, 139
Trapezium, 60
Trapezoid, 60
Traps, 110
Tread, 168
Trimmer joist, 107, 108, 123, 171
Tudor arch, 94, 246
Turning piece, 97
Turret roof, 324
Turrets, 265
Twist bit, 32
Twisting, 20

U

Upper floors, 105

V

Ventilation, 15
Vertical section, 74
Vessels, 11
Vestibule screen, 240

W

Wall brackets, 179
  hook, 183
  plates, 114
  string, 171
Walling pieces, 196
Wall piece, 188
Weather strip, 161
Western Hemlock, 21
Wet and dry thermometer, 18
White deal, 21
White fir, 21
Whitewood, 21
Width joints, 39
Winders, 177
Window board, 139
Windows, 254
  bay, 108, 139
  casement, 76, 136
  dormer, 258
  eyebrow, 261
Wire nail, 49
Wood grounds, 139
  plugs, 52, 136

Y

Yellow deal, 21
Yoke, 93